The **Politically Incorrect Guide**™ to

SCIENCE

The Politically Incorrect Guide™ to
SCIENCE

TOM BETHELL

Since 1947
REGNERY
PUBLISHING, INC.
An Eagle Publishing Company • Washington, DC

Library of Congress Cataloging-in-Publication Data

Bethell, Tom.
   The politically incorrect guide to science / Tom Bethell.
      p. cm.
   Includes bibliographical references and index.
   ISBN 0-89526-031-X
   1. Science—United States. I. Title.
   Q127.U5B48 2005
   509'.73—dc22

                              2005029108

Published in the United States by
Regnery Publishing, Inc.
One Massachusetts Avenue, NW
Washington, DC 20001
www.regnery.com

Distributed to the trade by
National Book Network
Lanham, MD 20706

Manufactured in the United States of America

10 9 8 7 6 5 4 3 2 1

Books are available in quantity for promotional or premium use. Write to Director of Special Sales, Regnery Publishing, Inc., One Massachusetts Avenue NW, Washington, DC 20001, for information on discounts and terms or call (202) 216-0600.

CONTENTS

Introduction: **The Lures of Politics**   v

Chapter 1:   **Global Warming**   1
        We love the Seventies
        Evidence for warming: watch a hockey game
        Profiles in courage

Chapter 2:   **Yes, More Nukes**   19
        The new alchemy
        Ignorance, fear, and the search for virtue
        A "green" civil war
        Piddle power

Chapter 3:   **Good Vibes: The Virtue of Radiation**   39
        Background radiation
        Chernobyl
        Plutonium myth
        Catching some (gamma) rays

Chapter 4:   **"Good Chemistry"**   57
        The Comeback Kid: Dr. Edward Calabrese
        The dioxin panic
        Studying dioxin

Chapter 5:   **The DDT Ban**   73
        Silent but deadly: Rachel Carson
        Politicized science
        Walking on eggshells
        EPA ban
        Africa pays
        The learning curve
        How to rescind the ban

Chapter 6: **Biodiversity and Endangered Species** 87

How many species, how many extinguished?

Political science

Actual cases

Political taxonomy

Issues of property

Chapter 7: **African AIDS: A Political Epidemic** 105

AIDS redefined

False positive

Life and death

Quiet population explosion

Post-colonial health

Chapter 8: **The Folly of Dolly: Cloning and Its Discontents** 123

Chapter 9: **The Stem Cell Challenge to Bioengineering** 131

Chapter 10: **A Map to Nowhere** 147

Genome is decoded—again

Not in the genes? We always knew that . . .

Worthless genes

The unsolved problem

Gene therapy hits kids with SCID

Venter hits jackpot, and Polynesia

Chapter 11: **The Great Cancer Error** 165

Chapter 12: **The Abiding Myths: Flat Earth and Warfare between Science and Religion** 181

The true "Flat Earthers"

The warfare myth

Another side to the story

Chapter 13: **By Chance, or by Design?** 199

The origin of a theory: Darwinism and its weakness

Irreducible complexity

Contents

Chapter 14: **Evolution: The Missing Evidence**   **215**

"There are no half-bats."

Haeckel's embryos

Peppered moths

The tree of life

"Building blocks" . . . in a flask

Darwin's finches

"On the tendency of varieties to depart
indefinitely from the original type"

Final Thoughts   **237**

Notes   245

Index   259

Introduction

THE LURES OF POLITICS

Scientists seem to enjoy a measure of immunity. Scrutiny is tolerated, but preferably from within their own ranks. But science is subdivided into a thousand fields, and therein lies a problem. Experts hate to challenge one another, just as doctors do. Often, for a specialist in one field to appreciate what others are saying, careful study must be undertaken. Time is always too short. Outsiders will fear to enter others' fields in anything other than a deferential spirit. So challenge and disagreement rarely arise. The priesthood of science is undisturbed, and that is the way they like it.

But science surely has become politicized, and if scientists won't blow the whistle on each other, who will? Journalists, I believe, need to become more involved. Not just in reporting science, but in assessing it more critically.

Journalists are generalists, often quite adept at acquiring basic knowledge in a new field. But they, too, can be reluctant to challenge experts. This is especially true in the medical field. Senior officials from the National Institutes of Health have been known to call television producers and tell them that certain views, if publicized, could endanger the health of the nation. Journalists sometimes think it's downright *unsafe* to question the experts. Actually, it's unsafe not to.

The old and admirable refrain, often heard from the newsrooms at the time of Watergate, was: "Don't accept government handouts." But that tends to be forgotten when medical science is concerned. I once asked a journalist why he so uncritically printed government handouts about AIDS. "I don't have a license to practice medicine," he said.

In the political arena, including intelligence and foreign policy, journalists take greater liberties. Yet this wasn't always the case. Here, too, they received phone calls warning about national security. Then, thirty-five years ago, leading editors decided to form their own judgments. The context was the Vietnam War. The Nixon administration tried to prevent publication of the Pentagon Papers, a critical history of the war. Judicial restraining orders were issued, but eventually the public's "right to know" prevailed.

The *New York Times* published the material, and then other newspapers followed suit. And we were better off for it. The role that journalists play when they *challenge* the government is beneficial, whatever the bias in reporting. The problem is that all too often they don't challenge government policy, but promote it, even when the government's mission is obviously self-interested. The real danger is unchecked government power.

National reporters regard the Pentagon, the State Department, and the CIA as fair game. Advanced degrees in international relations are not expected, or even particularly respected. Journalists cultivate sources within the agencies; in effect, they use the power of publicity to play a role in the development of policy. But when it comes to science they are more deferential. Scrutiny of scientific claims has lagged way behind.

The budget of the National Institutes of Health has doubled under President Bush, and because of post–September 11 security concerns, a fence was built around the huge facility. It is a literal fence, but also a symbolic one: Do not disturb! A white-coated priesthood is at work. Cures are at hand. It is taxpayer-funded, mind you, but journalists are intimidated. If only the skepticism with which they view the uniformed officers of the Department of Defense could be extended to the uniformed officers of the Public Health Service.

Jonathan Fishbein was a whistle-blower at NIH. He took exception to the cover-up of indications that a drug destined for African children was dangerous. But he received little attention from the major media (although the Associated Press was an exception). A few stories appeared, but they were buried, and soon forgotten. The *Washington Post*, so effective at the time of Watergate, sometimes seems to regard the thousands of employees at NIH headquarters in Bethesda, Maryland, as an important component of the paper's subscriber base.

This is one reason why a *Politically Incorrect Guide*™ is needed. In the realm of science, Woodward and Bernstein have not been on the case. And without oversight, the professionals can sometimes get away with murder.

Briefly, it's worth comparing the treatment of medical science with "the dismal science," as Thomas Carlyle called economics. Expertise in economics often receives short shrift, and economist jokes are two-a-penny. The experts don't get off easily just because they have the right credentials. And we sense that that is healthy.

Why is this? In the nineteenth century, economics was known as "political economy," and we recognize that that was its proper title. Economics really is more political than scientific. And politics is a field where journalists do not fear to tread. For that we are grateful (even if Karl Marx was a journalist).

For decades, as Michael Crichton points out, science was assumed to be above politics. It deals with facts, after all, not opinions or judgments. Facts are verified experimentally, and experiments can be repeated. Science is a self-correcting field. (True—in the long run.) Politics, in contrast, is an arena of contending values.

But it turned out that science could easily be politicized. The most important reason is this: Often, there is a great deal of uncertainty as to

The Other Political Science

"Science has been the great intellectual adventure of our age, and a great hope for our troubled and restless world. But I did not expect science merely to extend lifespan, feed the hungry, cure disease, and shrink the world with jets and cell phones. I also expected science to banish the evils of human thought—prejudice and superstition, irrational beliefs and false fears. I expected science to be, in Carl Sagan's memorable phrase, 'a candle in a demon-haunted world.'

"And here, I am not so pleased with the impact of science. Rather than serving as a cleansing force, science has in some instances been seduced by the more ancient lures of politics and publicity. Some of the demons that haunt our world in recent years are invented by scientists. The world has not benefited from permitting these demons to escape free."

Michael Crichton,
Michelin Lecture at Caltech, January 17, 2005

what the facts are. Preferences can then be substituted for facts, and that can happen inconspicuously.

Global warming is a good example today. It is often said: If we don't know whether to take an umbrella to work, how can we predict the weather a hundred years off? Some of those who are the most vocal about warming today were talking about global cooling twenty-five years ago. If the globe is warming, is mankind responsible, or is the sun? Inevitably, amidst such uncertainties, the struggle to establish the relevant facts turned into a political struggle.

Many do now realize that. In consequence, something resembling a real debate on the subject of global warming has broken out. It is widely accepted that climate science itself is uncertain, and that some of the alarmists have a political agenda—to restrain U.S. economic growth, for example. There has also been some excellent reporting on the subject.

The result is that the issue has lost some of its potency. Once the warmists are seen not as impartial scientists but as political advocates, they lose credibility. The same is true of other environmental issues—endangered species, or clean air. The political agenda behind them has become more conspicuous. Meanwhile, their advocates have become bureaucratically entrenched and perhaps impossible to dislodge.

All science based on dire warnings about the future should be suspect, and all such science is almost by definition politicized—if only because democracy as presently constituted responds with undue haste to any and all claims of crisis. In 1798, in England, the economist Thomas Robert Malthus—a dismal fellow, to be sure—warned that the population was expanding more rapidly than the food supply. Parliament did precisely nothing, and was right not to. Malthus's mumbo-jumbo about arithmetical and geometric rates of increase confounded his critics for years, but it was all based on false premises. He was wrong, even if he did sound scientific.

Thirty years ago, all over the Western world, Malthusian scares about overpopulation resonated anew. Now it was a world crisis! Biologist Paul Ehrlich foresaw millions of American dying of starvation (obesity would have been a better prognosis). The U.S. shipped billions of condoms abroad—seven billion by 1990, according to one estimate. Now, however, we are beginning to hear about the potential problems of underpopulation.

In 2000, a "pandemic" of HIV/AIDS in sub-Saharan Africa was thought so severe that Vice President Al Gore and Secretary of State Madeleine Albright took the issue to the UN Security Council. It was heterosexually transmitted, we were told, so condoms were still needed. Today, sub-Saharan Africa has the most rapidly growing population in the world. (But more than ever, you can be sure, condoms will be needed.)

Future facts are unknown, in short, and uncertainty translates into opportunity for those who seek to politicize science.

All government agencies face essentially the same incentives. They benefit by persuading us that we can't get along without them. That may well be true of long-established departments such as Defense, State, and Justice, but it's less clearly true of agencies that were created recently, and in response to a supposed emergency. (The EPA comes to mind.) All such agencies deploy the same publicity campaign: *"The problem is even greater than we thought, but don't worry, we are making headway in solving it. So increase our budget—now!"*

Journalists should be suspicious of all such campaigns, whether they aim to arouse our fears or our hopes. Take the Human Genome Project. It was a government project from the beginning—"congressionally driven," as science god James Watson said. And he meant that as a compliment. Great medical benefits were promised, but nothing has materialized, and probably won't. Yet apart from some (justifiable) criticism from the Left, deploring the project's "determinist" ideology, the genome project has received nothing but hosannas from the mainstream press.

At the level of basic science, the genome project may in the end teach us a lot, if only because it has revealed the depth of our ignorance. The concept of the gene itself will probably have to be revised, and textbooks rewritten. An excellent article in *Harper's*, "The Spurious Foundation of Genetic Engineering," spelled out many of the details. But the original promise has not been fulfilled.[1]

Now the plan is to move on to a "cancer genome project." What we can say for sure is that it will create full-time employment for statisticians, and that Congress will rush the money over to the NIH. But it is unlikely to advance our understanding of cancer.

When a profitable opportunity looms and then private investment turns south, it is always a revealing indicator. In the case of genomics, the "business model" was said to be inappropriate, but the miscalculation was more fundamental than that. The science itself was askew. Something similar may be happening with stem cell research. Federal funding is restricted, but the research itself is legal. Yet if the medical promise is so great, why is the federal government so essential? Do venture capitalists know something the headline writers don't?

Journalists sometimes overlook these questions. Perhaps the reason is that, in the case of stem cells, the story has already been framed as a contest between promising science and reactionary ethics. Scientists certainly prefer it that way. The unsolved scientific difficulties scarcely get into the headlines at all.

Deep Throat: "Follow the Money."

In the film *All The President's Men*, Deep Throat tells Bob Woodward of the *Washington Post* that the key to unlocking the Watergate scandal is "Follow the money." In August 2005, the *Washington Post* reported that there was turmoil at the National Institutes of Health over employees investing in drug and biotechnical companies. Maybe we should follow the money too and find out what scientists themselves believe will be the next medical breakthrough.

"Flooded with 1,300 comments by employees and threats of high-level defections, the head of the National Institutes of Health agreed yesterday to loosen some of the ethics rules he unveiled in February.

"Under the final regulations, about 200 senior staff members will be required to divest large stock holdings in drug and biotechnology companies, NIH director Elias A. Zerhouni said. That is far fewer than the 6,000 employees who would have had to divest under his original proposal to strengthen conflict-of-interest rules at the world's premier biomedical research agency.... Many warned that the broad divestment order would have severe economic repercussions and cause some top agency scientists to leave for more lucrative jobs. A handful made headlines through the spring with rumors of their impending departures because of the regulations. One prominent scientist from Duke University said the restrictions could keep him from accepting a job at the institute."

Ceci Connolly, "Director of NIH Agrees to Loosen Ethics Rules," *Washington Post*, August 26, 2005

Other issues are "political" in a different way. Take cancer research. In Chapter 11 I argue that for three decades the National Cancer Institute has pursued an erroneous theory of cancer's origins—the gene-mutation theory. It's not that the scientists involved—the great majority of cancer researchers—have adopted a theory for political reasons. They haven't. If the argument here is correct, the underlying problem is created by government funding itself, which has obstructed the pursuit of alternative theories.

A government strategy of funding conflicting theories would look hit-or-miss: not much better than trial and error. Plainly, most research money would be "wasted," and politicians don't like that because they get the blame. Better to let the experts decide—form a consensus among themselves, form committees, and let them allocate the funds that way.

In contrast, private-sector research is trial and error, by its nature. Capital is invested in a wide range of ideas and approaches, and maybe only one will pay off. In the private sector it's called risk, not waste. The greatest scientific advances we have seen in recent decades, in the field of computer and information technology, did involve a great deal of risk and a great deal of "wasted" investment. But there was also tremendous progress.

Historically, the competition of theories has been the driving force behind scientific progress. Isolated individuals and private companies have been the most fruitful sources of this advance. And just as a competitive market system forces innovation into private enterprise, so the competition of theories drives science to investigate new approaches.

When all research eggs are in one basket, it's a different world. Competition may stagnate, or be eliminated entirely, if that is what the government decrees (as happened under Communism). When any single source of funding dominates, science will almost certainly become the handmaiden of politics. There is no recognition in our leading journals that this is a problem. *Science* magazine, for example, keeps a vigilant watch on government science spending, unhesitatingly equating "more" with better.

Government funding has also promoted the idea that a theory can be regarded as true if it enjoys enough support. There is certainly a consensus behind the gene mutation theory of cancer. Consensus discourages dissent, however. It is the enemy of science, just as it is the triumph of politics. A theory accepted by 99 percent of scientists may be wrong. Committees at the National Institutes of Health that decide which projects shall be funded are inevitably run by scientists who are at peace with the dominant theory. Changing the consensus on cancer will be an arduous task, like turning a supertanker with a broken rudder.

A Book You're Not Supposed to Read

Politicizing Science: The Alchemy of Policy-making by Michael Gough, ed.; Stanford, CA: Hoover Institution Press, 2003.

The theory of evolution is also supported by an overwhelming consensus. But is it true? The difficulty of knowing what the facts are (or were) is once again paramount. They took place hundreds of millions of years ago, give or take a zero or two, and physical decay has rendered those facts mostly unknowable. Fossils are sparse and difficult to interpret.

So we have few facts, but now the unknown lies in the past rather than the future. Fossils tell us that most organisms that once roamed the earth no longer do. From that a number of conclusions may follow, of which evolution is but one.

We are strongly inclined to substitute faith for uncertainty. Recently, Ben Adler of the *New Republic* asked a number of prominent conservatives if they "believe in evolution." Faith statements were duly elicited ("I do believe," "Of course," "Yes"). It was a strange question. No one seemed to notice that belief is more appropriately applied to religion than to science.

Chapter 1

GLOBAL WARMING

We have all heard the scenario. The world is poised for ecological disaster because man is polluting the atmosphere and heating up the earth. Global warming will melt the polar ice caps and cause the oceans to rise, submerging large parts of Miami, New York City, and other coastal cities. If you live in Manhattan, you'd better move to South Jersey, or better yet, Omaha, Nebraska.

You would think that with such predictions afoot, someone had been studying the data for a long time. At least, you would hope so. But global warming became the pet cause of environmentalists only in the late 1980s. Before then, some believed the earth was cooling, not warming. "The drop in food output could begin quite soon, perhaps only in ten years," *Newsweek* warned on April 28, 1975. "The resulting famines could be catastrophic." To stop global cooling, some experts proposed melting the Arctic ice cap! Now we are taught to fear exactly that. What is going on here?

According to the most reliable summaries of the earth's surface temperatures for the whole globe, which go back no further than 1861, there was a warming period in the first half of the twentieth century, lasting from about 1910 to 1940. That was followed by a cooling period from 1940 to 1975. Since 1975, we have experienced a slight warming trend.

Guess what?

⚬ Environmentalists not so long ago believed the earth was cooling.

⚬ The earth surface temperature data suggests that man-made greenhouse emissions have not been sufficient to increase global temperatures.

⚬ The Kyoto Treaty, which bound signatories to reducing greenhouse emissions, would have caused a depression in the United States.

The three periods combined give us a surface temperature increase of perhaps one degree Fahrenheit for the entire twentieth century.

But there is a problem. Satellite measurements of *atmospheric* temperatures do not agree with these surface readings. Satellite measurements began only in 1979, and they have shown no significant increase in atmospheric temperature in the last quarter century. Balloon readings did show an abrupt, one-time increase in 1976–1977. Since then, however, those temperatures seem to have stabilized.

Environmentalists believe that the twentieth-century warming was caused by human activity, primarily the burning of fossil fuels. Their combustion produces carbon dioxide—one of several "greenhouse gases." Methane is another. The argument is that their release into the atmosphere wraps the Earth in an invisible shroud. It makes the escape of heat into outer space slightly more difficult than its initial absorption by the Earth (from sunlight). This is the Greenhouse Effect. And the result is that the Earth warms.

The effect itself is not disputed by scientists, but whether man-made carbon-dioxide emissions have been sufficient to cause measurable

Peddling Fear

Stanford climatologist Stephen Schneider, winner of a MacArthur Fellow "genius" award in 1992, was quoted as saying: "We have to offer up scary scenarios, make simplified, dramatic statements, and make little mention of any doubts we might have. This 'double ethical bind' we frequently find ourselves in cannot be solved by any formula. Each of us has to decide what the right balance is between being effective and being honest. I hope that means being both."

Discover, October 1989

global temperature increases over the last thirty years is a matter of fierce debate. Carbon dioxide itself is a benign and essential substance. Without it, plants would not grow, and without plant life, animals could not live. The increase of carbon dioxide in the atmosphere therefore causes everything from plants to trees, forests to jungles, to grow more abundantly.

The surface data itself suggests that man-made carbon dioxide has *not* been sufficient to increase global temperatures. Consider the period 1940–1975, a time of considerable fossil fuel consumption. Coal-fired plants emitted smoke and fumes without any Green Party or environmental ministers to restrain them. Yet the Earth cooled slightly. Also, if man-made global warming is real, atmospheric *as well as* surface temperatures should have steadily increased. This has not happened. Increases were recorded only in the late 1970s, but these were probably caused by a solar anomaly, not by anything man was doing.

Where the temperature readings are made is vitally important. Within the United States an "urban heat island effect" has been identified. Build a tarmac runway near a weather station, and the nearby temperatures readings will increase. It all seems perfectly reasonable. But common sense is often in short supply when you are dealing with today's environmentalists. Meanwhile, Antarctica has been cooling even as Greenland is warming.

We love the Seventies

The first Earth Day was held in 1970, a nostalgic moment for today's environmentalists. Twenty-five million people participated, and Congress adjourned to "listen" to their constituents. In rapid succession Congress passed the Clean Air, Clean Water, and Endangered Species Acts. The Environmental Protection Agency was hurriedly brought onstage. By 1980, Jimmy Carter's "Global 2000" report forecast (pessimistically) global conditions expected to prevail at the end of the millennium. But the report failed to mention any warming trends.

By 1990, global warming (along with the claimed loss of "biodiversity," caused by human destruction of habitat) had become the most popular issues for environmentalists. In 1992, representatives from 160 nations met in Rio de Janeiro for the Earth Summit. The mood was anti-American, with images of "Uncle Grubby" substituted for Uncle Sam. President Bush (the elder) refused to sign the biodiversity treaty, but he did sign a treaty on climate change. Signatories agreed to reduce their emission of carbon dioxide.

The details of which countries would have to comply were worked out in Kyoto, Japan, five years later. Greenhouse-gas emissions were to be reduced to below their 1990 levels by 2012. That was the Kyoto Protocol. But President Clinton did not submit the treaty to the Senate for ratification; he knew it would never pass. Almost everyone knew that America was the principal target of the treaty. The 1990 date had been carefully chosen. Emissions in Germany and the Soviet Union were still high then; Germany had just swallowed up East Germany, then using inefficient coal-fired plants. After these plants were modernized, Germany's emissions dropped, so the demand that they be reduced below 1990 levels had already been met.

The same was true for the Soviet Union. After its collapse, in 1991, economic activity fell by about one-third. Today, Russia is still below its old emission levels. As for France, most of its electricity comes from nuclear power, which the environmentalists agree has no global warming effects but has been demonized for other reasons.

Under the Kyoto protocol, U.S. emissions would have to be cut so much—perhaps by one-third—that economic depression would be the one sure result. Meanwhile, Third World countries are exempt; so are China and India. Australia, like the United States, has refused to ratify the treaty. Thirty-five countries, mostly in Europe, have agreed to reduce their CO_2 emissions. But there are no enforcement mechanisms. The potential for cheating is almost unlimited, and by the time the Kyoto Protocol went into effect, in February 2005, the principal irritation was that the main target, the United States, had dodged a bullet.

Fred Singer, an atmospheric physicist at George Mason University and founder of the Science and Environmental Policy Project, is a leading critic of the global warming scares. He says in defense of the U.S. anti-Kyoto position: "We're being asked to buy an insurance policy against a risk that is very small, if at all, and pay a very heavy premium. We're being asked to reduce energy use, not just by a few percent but, according to the Kyoto Protocol, by about 35 percent within ten years. That means giving up one-third of all energy use, using one-third less electricity, throwing out one-third of all cars perhaps. It would be a huge dislocation of our economy, and it would hit people very hard, particularly people who can least afford it."

Meanwhile, the rhetoric, if not the globe, became more and more heated. The real threat is the rhetoric itself. Any unusual weather event may now be linked to climate change. Interviewed by movie star Leonardo DiCaprio in 2000, President Clinton said that if we do not change our ways "the polar ice caps will melt more rapidly, sea levels will rise...." The overall climate of North America could change, with "more flooding, more heat waves, more storms, more extreme weather events generally."

Within twenty-four hours of the Pacific tsunami in December 2004, *CBS Evening News*, citing unnamed "climate experts," displayed a

Collateral Damage

"No matter if the science is all phony, there are collateral environmental benefits... Climate change [provides] the greatest chance to bring about justice and equality in the world."

Christine Stewart, former Canadian Minister of the Environment, quoted by the *Calgary Herald*, December 14, 1998

graphic that had only the words "global warming" and "tsunamis." News anchor Dan Rather intoned: "Climate experts warned today that tsunamis could become more common around the world and more dangerous. They cite a number of factors, including a creeping rise in sea levels believed to come from global warming and growing populations along coastal areas."

Evidence for warming: watch a hockey game

The claim that the globe is warming depends on our knowledge of earlier temperatures. Since climate experts generally accept that throughout the twentieth century temperatures rose at most by one degree, that knowledge needs to be precise. But knowledge of ancient temperatures can be obtained only indirectly. Scientists depend on tree rings, boreholes, ice cores, and the skeletons of marine organisms deposited in the Sargasso Sea to decipher temperatures in earlier centuries.

Yet the graph that was most effective in persuading policymakers that scary things are happening has a horizontal axis covering a thousand years, and a vertical axis with temperature units separated by fractions of a degree. The temperature line is mostly horizontal, perhaps declining slightly for nine hundred years, and then abruptly heading up into a warmer range over the last one hundred years. The line is known as the "hockey stick," with the long handle representing the nine hundred years, and the blade the last hundred.

All warming scenarios, including the hockey stick, rely on mathematical "models" extrapolating from a vaguely known past to an unknown future. The UN's Intergovernmental Panel on Climate Change (IPCC), an advisory body that releases annual reports, boldly estimates a global temperature increase of five degrees Celsius in the twenty-first century and wants $200 billion dollars a year to prevent it. If planned remedies are installed, the UN panelists acknowledge that the temperature increase they foresee will merely be delayed by six years.

A powerful criticism of global warming scenarios comes from comparing data from different sources. For example, the temperature records from tree rings in the twentieth century can be compared with records from meteorological instruments. The comparison casts doubt on global warming because the data differ for recent decades when warming is alleged to have occurred. In the early part of the century instruments and tree rings yielded similar readings, but they begin to diverge significantly after 1970.

From 1970 on, the instruments show higher temperatures than tree rings. One plausible explanation is the urban heat-island effect. Many land-based thermometers are located in or near growing cities, where buildings, pavement, and industrial activity raise local air temperatures, sometimes by several degrees. Tree ring samples, in contrast, typically come from forested areas. It is likely, then, that the instrumental record, on which the hockey stick's "blade" is based, has an upward bias from local warming at atypical "heat islands." A National Research Council panel has agreed that the urban heat island effect is a serious and unresolved problem.

The "hockey stick" was first published in 1998 by climatologist Michael Mann of the University of Virginia. It was immediately used by the IPCC to promote the idea that we have an unprecedented crisis on our hands. But the chart also aroused suspicions, because for years there had been broad agreement among climatologists that in the second millennium AD, global temperatures were not as unvarying as Mann's chart implied. There had been ups and downs—periods of both warming and cooling. Beginning around 1000 AD, there was something called the Medieval Warm Period, which persisted until a period known as the "Little Ice Age" took hold in the fourteenth and fifteenth centuries. Both periods lasted for several hundred years.

The warmer period, accompanied by a flowering of prosperity, knowledge, and art in Europe, seems to have been wholly beneficial. Agricultural yields increased along with the temperature. Marshes and swamps—today they would be called wetlands—dried up, removing the breeding grounds of mosquitoes that spread malaria. Infant mortality fell; the population grew. From 1100 to 1300 AD, the population of Europe increased from about forty to sixty million.

One sign of the warming trend was the settlement of Greenland by Vikings from Iceland. They reached a peak of prosperity in the twelfth and thirteenth centuries, but began experiencing difficulties in the late fourteenth century, with the onset of the Little Ice Age. The settlements finally perished in the fifteenth century.

A recent review of papers reconstructing the climate from tree rings found seventy-nine studies showing "periods of at least fifty years which were warmer than any fifty-year period in the twentieth century," according Willie Soon and Sally Baliunas of Harvard.[1] The warm period has been recognized in climate textbooks for decades. It was an obvious embarrassment to those claiming that the twentieth-century warming was a true anomaly. The earlier changes occurred when fossil fuel consumption could hardly have been the culprit and prove that warming occurs without human action.

Profiles in courage

Consider, in this context, the experience of Dr. David Deming, an assistant professor at the University of Oklahoma's College of Geosciences. In 1995, he published a paper in the journal *Science*, reviewing the evidence showing that borehole temperature data recorded a warming of about one degree Celsius in North America over the last 100 to 150 years. Deming subsequently learned:

> With the publication of the article in *Science*, I gained significant credibility in the community of scientists working on

climate change. They thought I was one of them, someone who would pervert science in the service of social and political causes. So one of them let his guard down. A major person working in the area of climate change and global warming sent me an astonishing e-mail that said, "We have to get rid of the Medieval Warm Period."

Mann was already working on it. Whether he intended to is another question, but the hockey stick eliminated that pesky Medieval Warm Period. The twentieth century was going to be warmest, regardless of the data.

At the same time, Deming had his first encounter with the agenda-driven news media. He had ended his *Science* article with what he thought was an uncontroversial statement: "A cause and effect relationship between anthropogenic activities and climatic warming cannot be demonstrated unambiguously at the present time." Simply stated, the evidence didn't warrant the conclusion that recent warming had been caused by man.

A reporter from National Public Radio then called him to discuss the *Science* article. But it turned out that he only wanted to know about that last sentence. "Did you really intend to imply that the warming in North America may have been due to natural variability?" he asked.

"Yes," Deming replied.

"Well, then, I guess we have no story," the NPR man replied. "That's not what people are interested in. People are only interested if the warming is due to human activities. Goodbye." And he hung up. Deming now knew how the media intentionally screen what the public hears and have their conclusions firmly in mind beforehand.[2]

Once disciples of global warming were able to convince governments and media that we are experiencing something unprecedented, it was an easy step to claim that we also face a catastrophe. They had a free hand to extrapolate from minor (and beneficial) warming data and argue that temperatures will keep going up until the coastal cities flood. But voices of reason have begun to question all this.

A Toronto minerals consultant named Stephen McIntyre, who has no credentials as a climatologist, has successfully challenged the hockey stick. He spent two years and $5,000 of his own money trying to uncover Michael Mann's methods. Mann did at first give him some information, but then cut him off, saying he didn't have time to respond to "every frivolous note" from nonscientists.

McIntyre was then joined by another Canadian, an economist at the University of Guelph named Ross McKitrick. In 2003, they published an article critical of the hockey stick, claiming that Mann "used flawed methods that yield meaningless results."

In response, Mann published a rebuttal, and revealed some new information that had not at first appeared. His original article had been published in the prestigious British journal *Nature*, which then had to publish a partial correction based on this new information. McIntyre

Original Sin

"[T]he burning of fossil fuels (a concomitant of economic growth and rising living standards) is the secular counterpart of man's Original Sin. If only we would repent and sin no more, mankind's actions could end the threat of further global warming. By implication, the cost, which is never fully examined, is bearable. So far the evidence is not convincing. It is notable that thirteen of the fifteen older members of the European Union have failed to achieve their quotas under the Kyoto accord—despite the relatively slow growth of the European economies."

James Schlesinger, *Wall Street Journal*, August 13, 2005.
Schlesinger was the first Secretary of Energy (1977–79).

believes that more errors remain concealed. But he has been frustrated, because he still doesn't know the formula Mann used to generate his graph. The good professor has refused to release it. A *Wall Street Journal* reporter doggedly pursued the matter and contacted Mann. He told the reporter: "Giving them the algorithm would be giving in to the intimidation tactics that these people are engaged in."[3] It's hard to imagine other branches of the sciences, or the credibility of scientists themselves, remaining unchallenged for refusing to divulge a formula.

Another voice of reason, Francis Zwiers, a statistician with Environment Canada, a government agency, finds that Mann's statistical method "preferentially produces hockey sticks when there are none in the data." He is backed up by a mainstream scientist from Germany, Hans von Storch, who accepts that Mann's technique could sharply underestimate past temperature swings. (Dr. von Storch has said he faced pressure from colleagues who feared that skeptics could misuse his results. The tendency in climate science is to "make only comments that are politically correct," he says.) Also, new research from Stockholm University on historical temperatures suggests past fluctuations were nearly twice as great as the hockey stick shows.

Hans von Storch has said:

> The pattern is always the same: the significance of individual events is processed to suit the media and cleverly dramatized; when prognoses for the future are cited, among all the possible scenarios it is regularly the one with the highest rates of increase in greenhouse gas emissions—and thus with the most drastic climatic consequences—that is chosen. Equally plausible variations with significantly lower emission increases go unmentioned.
>
> Whom does this serve? It is assumed that fear can motivate listeners, but it is forgotten that it mobilizes them only in the

short term. . . . Each successive recent claim about the future of the climate and of the planet must be ever more dramatic than the previous one. Once apocalyptic heat waves have been predicted, the climate-based extinction of animal species no longer attracts attention. Time to move on to the reversal of the Gulf Stream. Thus there arises a spiral of exaggeration. Each individual step may appear to be harmless; in total, however, the knowledge about climate, climate fluctuations, climate change and climatic effects that is transferred to the public becomes dramatically distorted.

Sadly, the mechanisms for correction within science itself have failed. Within the sciences, openly expressed doubts about the current evidence for climatic catastrophe are often seen as inconvenient, because they damage the "good cause," . . . The incremental dramatization comes to be accepted, while any correction of the exaggeration is regarded as dangerous, because it is politically inopportune. Doubts are not made public; rather, people are led to believe in a solid edifice of knowledge that needs only to be completed at the outer edges.[4]

Mann now concedes it is plausible that past temperature variations may have been larger than thought. The issue deserves further investigation but must not be overshadowed by political issues, he said, adding we don't really need the hockey stick anyway. "The contrarians would have us believe that the entire argument of anthropogenic climate change rests on our hockey-stick construction," he says. "But in fact some of the most compelling evidence has absolutely nothing to do with it, and has been around much longer than our curve."

But *Nature* recently noted that "many climate researchers believe it was premature of the International Panel on Climate Change to give the visually suggestive curve so much prominence." Fred Singer of George

Mason University says that, in light of the new information, "the hockey stick is dead." For his efforts in challenging the disciples of global warming, Singer was identified as a "naysayer" and given an inaugural "Flat Earth Award" by the so-called Green House Network. He is proud of the award and publicized it.

In *State of Fear*, Michael Crichton also emerged as an unexpected yet powerful critic of global warming. He studied the subject for a couple of years before writing his book, to which he added a section entitled "Author's Thoughts" and an appendix. He compared global warming science to eugenics, and in a speech at Caltech in 2003 he compared it to the search for extraterrestrials (which is also based on bogus science, he says).

Crichton warned the Caltech students to be suspicious whenever they hear that any scientific conclusion is based on a consensus, as we have often been told is true of global warming. Consensus science, he said,

is an extremely pernicious development that ought to be stopped cold in its tracks. Historically, the claim of consensus has been the first refuge of scoundrels; it is a way to avoid debate by claiming that the matter is already settled.

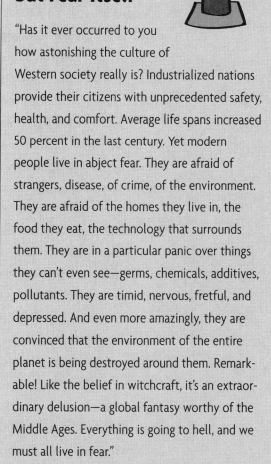

Nothing to Fear but Fear Itself

"Has it ever occurred to you how astonishing the culture of Western society really is? Industrialized nations provide their citizens with unprecedented safety, health, and comfort. Average life spans increased 50 percent in the last century. Yet modern people live in abject fear. They are afraid of strangers, disease, of crime, of the environment. They are afraid of the homes they live in, the food they eat, the technology that surrounds them. They are in a particular panic over things they can't even see—germs, chemicals, additives, pollutants. They are timid, nervous, fretful, and depressed. And even more amazingly, they are convinced that the environment of the entire planet is being destroyed around them. Remarkable! Like the belief in witchcraft, it's an extraordinary delusion—a global fantasy worthy of the Middle Ages. Everything is going to hell, and we must all live in fear."

Michael Crichton,
State of Fear; New York: HarperCollins, 2004, 455.

Whenever you hear the consensus of scientists agrees on something or other, reach for your wallet, because you're being had.

There may be some warming as a part of a natural trend that began about 1850 "as we emerged from a 400-year cold spell known as the Little Ice Age," Crichton said. But "no one knows how much of the present trend might be natural or how much man-made."

As a young man, Crichton studied at Harvard Medical School and the Salk Institute. He believes that "open and frank discussion" of global warming is being suppressed. One indication is that "so many of the outspoken critics of global warming are retired professors," Crichton said. They can speak freely because they are no longer seeking grants or facing colleagues "whose grant applications and career advancement may be jeopardized by their criticisms."

Another is that leading scientific journals "have taken strong editorial positions on the side of warming," which, Crichton said, "they have no business doing." He didn't identify the journals, but *Science* magazine itself is the leading offender. On most issues, but fortunately not all, *Scientific American* has likewise abandoned itself to political correctness.

The politicization of science was underscored recently when Dr. Naomi Oreskes of the University of California analyzed almost 1,000 papers on global warming published since the early 1990s. She concluded that 75 percent of them either explicitly or implicitly backed the consensus view, while none directly dissented from it.

Her study has been routinely cited by those demanding action on climate change.

But her conclusions raised suspicions. Other academics knew of many papers that dissented from the pro–global warming party line. They included Dr. Benny Peiser, a senior lecturer in the science faculty at Liverpool's John Moores University. He conducted his own analysis of the same papers and concluded that only one-third backed the consensus view, and only one percent did so explicitly. He submitted his findings to *Science* in January 2005 and was asked to edit his paper for publication. Then he was told that his results had been rejected because his points had been "widely dispersed on the Internet."

A Book You're Not Supposed to Read

Global Warming's Unfinished Debate by S. Fred Singer; Oakland, CA: Independent Institute, 1999.

Peiser replied that he had kept his findings strictly confidential. "It is simply not true that they have appeared elsewhere already," he told a London newspaper. *Science* then said that Peiser's research had been rejected "for a variety of reasons."

He is not the only academic whose work on this subject was rejected. Dennis Bray, a climate analyst in Germany, submitted results from an international study showing that fewer than one in ten climate scientists believed climate change is principally caused by human activity. Once again, *Science* refused to publish it. "They said it didn't fit with what they were intending to publish," Bray said.

The University of Alabama's Roy Spencer, a leading authority on satellite measurements of global temperatures, said: "It's pretty clear that the editorial board of *Science* is more interested in promoting papers that are pro–global warming. It's the news value that is most important." After his own team produced research casting doubt on man-made global warming, they were no longer sent papers by *Nature* and *Science* for review—despite being acknowledged as world leaders in the field.

As a result, Spencer said, flawed research is finding its way into the leading journals, while rebuttals are turned away. "Other scientists have had the same experience," he said. "The journals have a small set of reviewers who are pro–global warming."

This stifling of dissent and preoccupation with doomsday scenarios is bringing all climate research into disrepute. "There is a fear that any doubt will be used by politicians to avoid action," Benny Peiser said.

"But if political considerations dictate what gets published, it's all over for science."[5]

The issue of funding is critical. Scientists know only too well who is paying the piper, as Michael Crichton said. They know that "continued funding depends on delivering the results the funders desire." It's a variation on Gresham's Law: bad money chasing out good ideas. Environmentalists have become adept at de-legitimizing their opponents by saying they are "supported by industry," but studies funded by environmentalist organizations are "every bit as biased," Crichton added. They know their paymasters.

Myron Ebell, who works for the Competitive Enterprise Institute (CEI) in Washington, D.C., one of the few groups that critically examines global warming claims, says that lobbying for environmental causes is now a $1.6 billion industry. Skeptics like him are outnumbered by global warming advocates by perhaps 500 to 1 in the Washington, D.C., area. Yet CEI, hopelessly underfunded by comparison with such groups as the Sierra Club, is often characterized in the media as "industry supported."

The Unabomber Manifesto

"The two main tasks for the present are to promote social stress and instability in industrial society and to develop and propagate an ideology that opposes technology and the industrial system."

Theodore Kaczynski

Ebell says the real problem is that the environmental lobbyists have "everything going for them except the facts."[6]

What Michael Crichton and others can attest to is that the environmental movement has become a "special interest" like any other, with legislative goals and large budgets. Some leading environmentalists are even voicing criticism. In 2004, Michael Shellenberger and Ted Nordhaus wrote a 14,000-word essay that has since been widely circulated. Called "The Death of Environmentalism," it has "provoked a civil war among tree huggers," Nicholas D. Kristof wrote in the *New York Times*. It included some good lines of self-disparagement. Martin Luther King's "I have a dream" had found its counterpart in the environmentalists' "I have a nightmare." Shellenberger and Nordhaus wrote that this message was beginning to wear thin.[7]

In effect, it was a cry of anguish: why have we been unable to win on our top issues, especially global warming? They called it "the world's most serious ecological crisis," which (reverting to the nightmare) "may kill hundreds of millions of human beings over the next century."

They looked back fondly to their golden age in the early 1970s, when everything went their way. "It was then that the community's political strategy became defined around using science to define the problem as 'environmental'," they said.

"Using science" was a successful tactic. The rest of us were blinded by it for about twenty-five years. But the problem wasn't that the use of science had led environmentalists to propose unattractive "technical fixes," as Shellenberger and Nordhaus believed. The problem was that their science was never very good to begin with. And as its inadequacies became apparent, the scare tactics of the global-warming disciples became more and more conspicuous.

Chapter 2

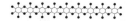

YES, MORE NUKES

In December 1953, President Dwight David Eisenhower gave his "atoms for peace" speech. He was addressing the young United Nations—an institute then looked to by many in the world for leadership. People still believed in experts (especially in science), in joint communiqués expressing goodwill among nations, and in the universal aspirations of people everywhere. If only the world's scientists and engineers had "adequate amounts of fissionable materials," Ike said, scientists would put it to good use. One special need was "to provide abundant electrical energy in the power-starved areas of the world." Governments should begin to make "joint contributions from their stockpiles of normal uranium and fissionable materials" to an agency of the United Nations. Not to worry—experts would control everything. In this way, the great powers would "serve the needs rather than the fears of mankind."[1]

The optimism now seems misplaced, but Eisenhower was right to argue that the time was ripe for nuclear power. In a few months, work began on the first commercial nuclear-power station, in Shippingport, Pennsylvania. A coal-powered plant had been planned, but the prospect of more air pollution was not welcome—the Pittsburgh area had experienced enough pollution from coal. So a utility, Duquesne Light, switched over to nuclear. In a well-publicized speech around the same time,

Guess what?

.•. Nuclear power, by far the cleanest and safest source of energy, was held up for twenty-five years after an accident in which no one was hurt.

.•. Global warming fears have given nuclear power a "green light" from (some) Greens.

.•. The level of emitted radiation that it is considered "safe" for a nuclear power plant is lower than the level of background radiation in Denver.

Lewis L. Strauss, the chairman of the Atomic Energy Commission, said that atomic energy would make electricity "too cheap to meter."[2]

There followed a time of rapid growth for nuclear energy. The future seemed rosy for this unlimited source of energy. Then came the 1960s and flower—not nuclear—power, when doubts about nuclear safety were publicized. A number of scientists, who should have known better, set out to scare the public and undermine confidence in the new technology.

Hollywood dutifully played its part. Neville Shute's 1959 movie *On the Beach*, starring Gregory Peck, was based on the fear that an all-out nuclear war would contaminate the entire world with lethal radiation. Fred Astaire's character says, "We're all doomed, you know. The whole silly, drunken, pathetic lot of us. Doomed by the air we're about to breathe." The film was acclaimed, but the science was wrong. While millions would die from the heat and blast, the residual radiation would generally be less intense "than people experience every day in Colorado, where it is rumored that people aren't dying in the streets," says physicist (and Colorado resident) Howard Hayden. The increase in background radiation "would die out rapidly anyway."[3]

The press was soon drawn to the scaremongering. The highly regarded TV commentator Edwin Newman said on NBC that as a result of the heat generated by nuclear power

"I Like Ike"

"The United States knows that if the fearful trend of atomic military buildup can be reversed, this greatest of destructive forces can be developed into a great boon, for the benefit of all mankind.

"The United States knows that peaceful power from atomic energy is no dream of the future. That capability, already proved, is here—now—today. Who can doubt, if the entire body of the world's scientists and engineers had adequate amounts of fissionable material with which to test and develop their ideas, that this capability would rapidly be transformed into universal, efficient, and economic usage?"

President Eisenhower's "Atoms for Peace" speech before the UN General Assembly, December 1953

plants, "by the end of the decade our rivers may have reached the boiling point."[4]

Then life imitated art. In 1979, Columbia Pictures released *The China Syndrome*, starring activist actress Jane Fonda. In the movie, the meltdown of a nuclear reactor core threatened to burn its way deep into the earth, "all the way to China." Two weeks later, there was a real nuclear accident, at Three Mile Island outside Harrisburg, Pennsylvania. This created "the most shocking synchronicity between real-life catastrophe and movie fiction ever to have occurred," according to Fonda. She and her then husband, the radical activist Tom Hayden, embarked on a fifty-two-city anti-nuke tour.[5]

Not surprisingly, fact and fiction became blurred in the public mind, and today, few people seem to realize that disaster was averted and no one in the plant or the Three Mile Island neighborhood was hurt. There was a small release of radioactivity, but the average dose received by a nearby resident was nine millirems—far less than received in a chest X-ray. Airline passengers on a single cross-country flight receive about two millirems from normal background radiation and cosmic rays. But the accident encouraged many people to think that the anti-nuclear activists had been right, and the construction of new nuclear power plants was halted.

Seven years later a Soviet reactor in Chernobyl, Ukraine, exploded and about fifty people died.[6] There were no confirmed deaths outside the plant itself. Radioactivity spread to the immediate area, and there were reports of thyroid cancer. But there was also an iodine deficiency—a risk factor for thyroid cancer—in the area. Today, the background level of radioactivity at Chernobyl is lower than that emitted by the granite of Grand Central Station.

For the anti-nuclear activists, who had been forced to resist nuclear power with every trick in the book, the combination of Three Mile Island, *The China Syndrome,* and later Chernobyl was a dream come true.

Activists took to the streets and commanded headlines. Comedian-activist Dick Gregory, known for hunger strikes, pledged that he would not eat solid food until all nuclear plants in the U.S. had been shut down. In the mind of the public, nuclear power had been discredited. Substitutes far less safe in terms of transportation hazards, accidents, and air pollution were adopted. Compare a thousand-megawatt coal-fired plant with a nuclear plant of the same capacity. Here is what each emits in the course of a year:

	Carbon Dioxide	Sulfur Dioxide	Nitrogen Oxides	Solid Waste
Coal	7 million tons	12,000 tons	20,000 tons	750,000 tons
Nuclear	None	None	None	50 tons

Meanwhile, the use of nuclear power continued without interruption in the U.S. Navy. Today 83 ships are equipped with 105 reactors, and there have been no accidents. These warships are welcomed at 150 foreign ports without encountering the local equivalents of Jane Fonda. On nuclear submarines, sailors work and sleep with their bunks only feet away from shielded reactors. They are allowed to receive an additional radiation dose of up to 5,000 millirems a year and report no ill effects.

Nuclear power always possessed one great vulnerability. Many people find it difficult to distinguish between nuclear power and nuclear weapons; atoms for peace or atoms for war, who knew the difference? The generic similarity is real. Both split atoms; both are radioactive. This has put nuclear power at the mercy of demagogues. On such technical issues, people are inclined to believe whatever they are told, and that is why the casual arousal of fears about all things nuclear was so irresponsible.

The new alchemy

What is nuclear power, and how does it work? Scientists long regarded the elements of nature as stable. Alchemists, Isaac Newton among them, had another idea, and believed that elements could be transmuted. Their dream was that "base" metals could be turned into gold. By the early twentieth century, this theory was turned inside out. Scientists found that some elements—particularly the heaviest—were not fixed after all. They transformed into other elements, which in turn broke down into still others. Transmutation lived.

The alchemists had been on to something—they just had it backwards. It turns out that lead was not the initiating element but, in most cases, the end product. And heat didn't need to be added—it was emitted. The philosopher's stone, the missing ingredient sought for many centuries by Newton and others, had been discovered. But now that it was real, it was itself transformed from the most desired to the most dreaded thing in the world. Its name was radioactivity.

Uranium is one of the mutable elements. It comes in two main varieties, or isotopes, designated U-235 and U-238. Over 99 percent of the uranium found in nature is of the latter, quiescent kind. U-235, on the other hand, is more volatile, and scientists realized that if they were able to separate out just that kind until it was over 90 percent "enriched," and a critical mass of it was brought together, it would explode.[7]

When one neutron is expelled from the uranium nucleus, it strikes the nucleus of an adjacent atom, and splits that atom, converting it into other elements. That releases a few more neutrons, which in turn strike further nuclei, and so on, in a chain reaction. BANG! Energy is released. Protons that were bound tightly together in the original configuration of the nucleus spring apart. That releases a lot of energy in a short time. The fission of one atom of uranium produces ten million times more energy than the combustion of an atom of carbon from coal.

But with a less enriched blend, say only 3.5 percent U-235 instead of 90 percent, the chain reaction still occurs, but more slowly. Instead of exploding, it fizzles. Heat is generated, and if this heating element is immersed in water, the water boils. That creates steam, which drives a turbine, which generates electricity. And that's all there is to it. A nuclear reactor is a great big kettle. Its internal element, consisting of uranium rods, heats up and boils water.

Nuclear power puts to good use qualities of matter that have existed all along but were discovered only a hundred years ago. The most difficult part of the whole process was the separation of the two isotopes of uranium, which (fortunately) is something that could be done only by technologically advanced societies.

One of the most misunderstood nuclear concepts is the half-life of radioactive material. Otherwise responsible publications often fail to inform their readers of this crucial point: radioactive elements with a short half-life are far more dangerous than those with a long half-life. Uranium-235, for example, has a half-life of 700 million years. Plutonium: 24,000 years. The two isotopes that have caused greatest concern are cesium-137 and strontium-90. Both have a half-life of about thirty years. Isotopes with very short half-lives never leave the reactor, so we don't have to worry about them.

When a radioactive substance has a short half-life, picture the Geiger counter clicking rapidly. A former acting Secretary of Energy put it like this: "Would you rather sit on a box of firecrackers if half will go off in the next week? Or a box in which half will go off in the next 24,000 years?"

Yet the long half-life of radioactive material is often cited as the most dreaded aspect of nuclear power, rendering contaminated sites uninhabitable for eons. That is false. The key variable is the rate at which particles radiating from a given volume of radioactive material strike the body. At a low rate they are harmless—they may even be beneficial. Natural background radiation subjects us all to a low-level bombardment anyway.

Unfortunately, government policy decrees that there is *no* safe level of radiation, and in so doing it has created a rationale for the anti-nuclear activists to oppose any and all man-made radiation, even when it is lower than that found naturally. In the Rocky Mountains, where uranium is abundant, natural radiation is relatively high. Bernard Cohen of the University of Pittsburgh offered to eat some plutonium if Ralph Nader, the activist's activist, would eat the same amount of caffeine. Nader, who had said that a pound of plutonium could cause eight billion cancers, refused the offer. Cohen later offered to eat plutonium on television, but producers and reporters were not interested. Yes, plutonium is dangerous, because you can make an atom bomb out of it, but its long half-life ensures that its radioactivity is not toxic to humans.

Books You're Not Supposed to Read

The Health Hazards of Not Going Nuclear by Petr Beckmann; Boulder, CO: Golem Press, 1977.

The Solar Fraud by Howard Hayden; Pueblo West, CO: Vales Lake Publishing, 2004.

Nuclear energy, set back for a generation by the arousal of unjustified fears, is a mature technology appropriate to our current state of development. It is in an expansionary phase elsewhere, notably China.[8] Around the world, about one hundred new nuclear power stations are on the books; in the U.S., three more are planned, of much improved design. No one in America has died as a result of nuclear power, but as Petr Beckmann argued years ago in *The Health Hazards of Not Going Nuclear*, tens—even hundreds—of thousands of people have died as a result of our growing dependence on coal.[9]

"What has Green anti-nuclear activism achieved since the seventies?" Peter Huber and Mark Mills asked in *City Journal*. "Not the reduction in demand for energy that it had hoped for, but a massive increase in the use of coal, which burns less clean than uranium."[10]

Today, 104 nuclear power plants are functioning in the U.S., providing 20 percent of the country's electricity. Sounds like a lot, but remember: 1,000 nuclear power plants had once been envisioned by the year 2000.

Ignorance, fear, and the search for virtue

Energy policy under the two Democratic presidents since 1976 has combined an equal measure of ignorance and irresponsibility. (Republican administrations, with only a few exceptions, tend to fear the political fallout from environmental activists and have approached the issue defensively.) Jimmy Carter donned a cardigan in the White House as a "lesson" in saving energy. The need to save energy was the "moral equivalent of war." Bill Clinton promised at Kyoto to reduce greenhouse gas emissions to 1990 levels without doing much of anything to help meet U.S. energy needs. In 2001, Energy Secretary Spencer Abraham said that the Clinton administration's energy policy had consisted of drawing up "a list of fuels it didn't like—nuclear energy, coal, hydropower, and oil—which together account for some 73 percent of America's energy supply. Some policy."[11]

New power plants and oil refineries were blocked; Alaska's untapped oil reserves declared off-limits lest the caribou be disturbed. Only the comfortably off could be so careless. The fantasy was that natural gas—the one conventional source of energy vaguely approved by the Green movement—would pick up the slack.

At about the same time, the Greens unleashed a publicity campaign on behalf of renewable energy. This includes "solar" and the energy generated by wind, "biomass" ("firewood" to you and me), and hydropower. As Howard Hayden wrote in *The Solar Fraud: Why Solar Energy Won't Run the World*, the hunt for renewable energy became a national fad.[12] As far as most environmentalists were concerned, "renewable" energy is the most virtuous kind because it doesn't pollute, doesn't contribute to global warming, doesn't quit, doesn't cost, doesn't run out. At least in theory.

It all began in the 1960s with the hippies, when it was thought that you could drop out of society and live in harmony with nature. You would be off the electrical grid, independent of the utilities. You could grow your own vegetables, hole up in your cabin, and even partake of a little home-made electricity. Then you could kick back and draw inspiration reading Thoreau by night. If the sun shone all day, the solar panel on the roof might even take the chill off the bath water, but not much more than that.

Eventually, however, the Greens realized that their vision wouldn't deliver anything more than piddle-power. Most people found living like hippies in the wilderness a step backwards for civilization. Some of the better-off activists, cushioned by trust funds, wouldn't have minded shutting down the whole country. They had the resources to withstand the political fallout. Others realized that if the lights went off, or if the price of gas rose too high, there would be a political backlash. The politicians might turn against the environmental activists instead of fearing them.

The smarter environmentalists realized that if they were to achieve their vision of renewable energy, backyard windmills just wouldn't get the job done. They would have to be scaled up considerably. So tax incentives were required. Politicians would be happy to deliver them— rather than face accusations of despoiling the environment. And once the tax incentives were in place, you could rely on private companies to come along and build whatever the environmentalists wanted.

Windmills became huge wind turbines, the size of loading cranes overlooking vast container ships. If you ever flew in to San Francisco, you might have seen in the East Bay rows of wind turbines perched on bare, usually brown hills. The area is known as Altamont Pass. These turbines are the older, smaller kind. The new wind machines are much bigger. But there is irony here, not often recognized. As the wind machines grew, they divided the activists and led to a "green" civil war. It was a conflict between groups that seemed likely to remain allies unto eternity—environmentalists and the lovers of renewable energy. And it led to a lawsuit.

The Union of Concerned Scaremongers

Since its founding in 1968, the ground zero of politicized science has been the Union of Concerned Scientists (UCS), in Cambridge, Massachusetts. Early on, UCS's signature issue was nuclear, beginning with opposition to anti-ballistic missiles in 1969, followed by opposition to nuclear testing, the strategic defense initiative, and support for nuclear disarmament.

Only in recent years has UCS adopted environmental issues. A few months before the 2004 election, it issued a politically slanted report accusing the Bush administration of having "manipulated and censored science to serve its political agenda." The issues cited included climate change, air pollution, and the supposedly unwise reliance on sexual abstinence rather than condoms.

The Concerned Scientists' interest in all things nuclear has not faded, and today they promote the idea that a nuclear power plant is a scary thing to have in your neighborhood. In a 2004 analysis, "Chernobyl on the Hudson," the organization irresponsibly concluded that the Indian Point nuclear power plant, "thirty-five miles upwind from Manhattan," poses "a severe threat to the entire New York metropolitan area."

Their analysis claimed to "find" that a "successful attack" by terrorists on the plant could produce "as many as 518,000 long-term deaths from cancer." This is two or three times the total number of deaths caused by the nuclear explosions over both Hiroshima and Nagasaki. (No one, not even UCS, thinks that a terrorist attack on a nuclear power plant could cause an atomic explosion.)

Scare stories sell magazines, and the mass media play along. *Time*'s "ARE THESE TOWERS SAFE? Why America's nuclear power plants are still so vulnerable to terrorist attack and how to make them safer," was a good example. Filling eight pages of the magazine, it postulated black-clad snipers with

.50-caliber rifles and bolt cutters, platter charges, gun-mounted lasers, infrared devices, electronic jammers, hand-drawn maps, and drawings of control panels. Then this:

"Once inside, the terrorists' hard work would be over. Then, surprisingly, would come the easy part: triggering a nuclear meltdown. They would spend a minute or two carefully flipping, disabling and breaking specific controls and shutting down pumps and operating key valves." This reactor meltdown "could" eventually kill "hundreds of thousands" of people.

How would the terrorists know what to do? "It would be a deadly sequence that they had mastered in advance from an accomplice who had probably worked in the control room of the reactor"—a "covert comrade working inside the plant."[13]

Oh. In these scenarios, the scaremongers are permitted to assume that all safeguards, no matter how elaborate, can be overcome by well-positioned traitors. That way, nothing can ever be safe. How about al Qaeda infiltrating the Secret Service?

Time's scenario was attributed to David Lochbaum, "a nuclear engineer who spent seventeen years working in reactors." In the next paragraph, however, he is described as "now a nuclear safety engineer with the Union of Concerned Scientists, a nuclear watchdog group." So the Concerned Scientists are still at it—arousing our concerns. They are better described as an anti-nuclear activist group.

A "green" civil war

Howard Hayden, who puts out a newsletter called "The Energy Advocate," had been predicting for decades that environmentalists would be the fiercest opponents of renewable energy. Still, it came as a surprise when the Center for Biological Diversity (CBD) sued a number of wind-power companies in California operating at Altamont Pass. The CBD is one of those perfectly conventional environmental groups, based in San Francisco, with dozens of blue-bloods in its upper echelons.

The first paragraph of the lawsuit is an eye-opener:

> This is a complaint to recover restitution from defendants for their past wanton killing of many thousands of protected birds, including thousands of raptors such as Golden Eagles, Red-Tailed Hawks, American Kestrels, falcons and owls. These killings are in flagrant violation of the criminal prohibitions of numerous provisions of the California Fish and Game code, the federal Bald Eagle and Golden Eagle Protection Act, and the federal Migratory Bird Treaty Act. Defendants have killed these magnificent raptors and other birds as a regular and continuing part of the process of generating electricity using thousands of small, obsolete wind turbine generators owned and/or operated by the defendants or entities they control at Altamont Pass in Alameda and Contra Costa Counties, California.[14]

Over 5,000 wind turbines operate in the Altamont Pass, and, according to the complaint, "they have killed tens of thousands of birds, including between 17,000 and 26,000 raptors—more than a thousand Golden Eagles, thousands of hawks, and thousands of other raptors."

More than a thousand Golden Eagles! Thousands of hawks! Remember what happened in Manhattan when that cruel co-op removed the nest of just one hawk, Pale Male? And these magnificent birds in California are being killed, in the thousands, by private, profit-seeking corporations who take the electricity and sell it to public utilities! Why aren't the guilty parties in prison already?

No one seems quite sure why these birds fly into the turbines, incidentally. We do know that eagles and other raptors have terrific eyesight.

They can spot a mouse a mile away. Hayden said that the seemingly slow rotation rate of the blades is deceptive.

> The tip speed of a wind turbine is approximately six times the wind speed—90 mph in a 15-mph wind—independent of the diameter of the wind turbine. What looks like a big fan making lazy circles in the sky (because of its low rpm) is actually three blades moving at high speed. So raptors see a blade move across their field of view and then disappear. They fly into the void only to be clobbered if they don't pass through the 6-to-10-foot gap by the time the next blade comes by. It gives new meaning to getting whacked.[15]

The lawsuit blames the "obsolete, first generation machines," some of which were installed twenty years ago. The latest wind turbines have blades much higher off the ground and generate more electricity. They are considered less deadly to birds "on a per-kilowatt basis." A single modern turbine can replace twenty or more of the older ones, the lawsuit claims.

So now they are the size of the Statue of Liberty. Some are to be tried out in Massachusetts, but they, too, are dividing the activists. There's a big project under way five miles off Cape Cod, in the hallowed waters of Nantucket Sound. One hundred and thirty turbines (over 400 feet high) will in theory be able to supply 75 percent of Cape Cod's electricity. But the Kennedy family worries the turbines twirling away on the horizon will spoil the view. And Robert Kennedy, Jr., environmentalist, is against the whole idea: He imagines that visitors want to see "what the Pilgrims saw when they landed on Plymouth Rock."

Walter Cronkite, once billed as the most trusted man in America, was opposed to the scheme in the beginning, but changed his mind. He actually thinks that Nantucket Sound "is a waste area, really." It's so shallow that "nobody would sail through it," says the old salt. The early

A Good Headline Is Worth a Thousand Words

Headlines have more influence than editorials. Consider nuclear waste. Its disposal is an important and unresolved issue. Many citizens, drenched in anti-nuke propaganda, think that a nuclear-waste storage site even a hundred miles away is the local equivalent of a death star.

Nevada's Yucca Mountain, almost a hundred miles from Las Vegas, was supposed to be ready by 1998 as the nation's main waste-storage site. But Nevada's politicians objected, and that gave them leverage to extract goodies from the feds and good headlines from the press.

A local newspaper reported it this way: "State Nuclear Projects Agency Chief Bob Loux said based on the EPA's estimates there will be 10 million cancer deaths over 1 million years that result from storing highly radioactive spent fuel in the mountain, 100 miles northwest of Las Vegas."

How did the *Nevada Review Journal* headline this story?

"State Official Says 10 Million Cancer Deaths Would Be Acceptable Under Safety Standard"[16]

Why do Nevadans oppose Yucca Mountain? Now we know! A "state official" thinks the facility might cause ten million acceptable cancer deaths. By the way, he owes his job to the influence of Senator Harry Reid, who finds it is easier to create jobs in Nevada by raising scares and keeping the facility out of his state than by being a good citizen and letting it in.

opposition, his included, was "almost hysterical." The most trusted American can't be trusted on this issue, at least among the activists.[17]

The Cape Wind Project is not yet under construction and it's an interesting question which side will prevail. The betting there is that Cronkite shrewdly switched to the winning side. The project will no doubt go forward, but some skeptics guess that maintenance will turn out to be a headache, and that ever-increasing subsidies will be needed to keep Cape Cod a well-lit place.

Remember, the argument is that the older, too-small turbines pose a threat to raptors because there isn't enough clearance between the blade tip and the ground. The new and improved turbines have a much larger clearance, which will save the birds. But then came this worrisome news from the *Washington Post*:

Researchers Alarmed by Bat Deaths from Wind Turbines

Now bats. Aren't they supposed to have pretty good built-in sonar? This was in Appalachia, where turbines the size of huge construction cranes rise 350 feet above the West Virginia mountains—well above the tree canopy. Once again, researchers are said to be "baffled," uncertain whether bats are "attracted to the spinning blades," or if their sonar, which allows them to avoid trees and even to catch mosquitoes in mid-air at the dead of night, "fails to detect the turbines." Many thousands of dead bats have been found, "some with battered wings and bloodied faces."[18]

The deaths "appear to violate no federal laws," says the U.S. Fish and Wildlife Service. But we do know that bats perform a useful service, gobbling up tons of mosquitoes and other unfriendly insects. Waiting in the wings is a group called Bat Conservation International, in Austin, Texas. Their leader is already talking about "unsustainable kill rates," so the renewable energy folk will probably be hearing from them soon.

Piddle power

One thing we do know about renewable energy is that it is not getting the job done. In 1979, President Carter called for a "national commitment to solar energy," with the goal of producing 20 percent of the nation's energy from so-called renewables by the year 2000. Remarkably, the renewables' actual contribution to the energy pie declined from 1980 to 2000, in percentage terms. And that was despite the new tax credits and subsidies.

Renewables contributed a mere 5.9 percent of the nation's energy in 2002. But that includes hydroelectricity, by far the greatest contributor. Environmentalists hate it, because it involves damming rivers, flooding scenic areas, and obstructing the passage of salmon. Clinton's EPA head, Carol Browner, refused even to dignify it as a renewable energy source. The environmentalists would like to tear down as many dams as they can.

Leave out hydropower and firewood, Howard Hayden says, and the residual "high-tech" sources, meaning photovoltaics ("solar") and wind, contribute a mere 0.19 percent of total U.S. energy needs; hopeless, in other words.[19]

It's hard to disguise these numbers, though the media try. For example, wind is said to be the most rapidly increasing source of energy. (Yes, but from a tiny base.) In its bat-kill story the *Washington Post* admitted that the wind-turbine industry "provided nearly 17 billion kilowatt hours, enough to serve some 1.6 million households—less than 1 percent of the country's electricity production."

Don't even ask about the environmental impact of generating commercial electricity from sunlight, because the environmentalists cringe when the subject is brought up—and rightly so. There's a demonstration project called Solar Two in the Mojave Desert, and that is pretty much where to put these things if solar has any chance at all. It may seem obvious, but you need a lot of sun for solar energy, which means that it works quite well at mid-day in Barstow, California, in the western Mojave.

The basic problem with wind and solar is that they are already "dilute" sources of energy. A magnifying glass can "concentrate" sunlight on a spot of paper, but that is only one spot. Same problem with photovoltaic cells: Small photovoltaics can power an electronic calculator, but not even the dimmest light bulb.

For an installation of solar reflectors to produce as much power as a typical nuclear plant in a year, Hayden writes, "it would have to cover 127 square miles." In other words, an area twice the size of Washington, D.C., would have to be covered with movable mirrors. And to maintain their efficiency, these mirrors must be washed every few days. Oh, and there has to be a natural gas back-up system to keep the therminol (fluid) bubbling when it's cloudy, or when the sun has set.

Tax credits determine all business decisions in the renewable field, and to qualify for them, natural gas can supply no more than 25 percent of the energy generated. "And that's about how much they do use," Howard Hayden says.

How much land does commercial wind power need? Imagine a mile-wide swath of windmills extending all the way from San Francisco to Los Angeles (400 miles). That land area is what would be required to produce as much power around the clock as one large coal, natural gas, or nuclear power station, which normally occupy one square kilometer of land.

As to the lawsuit, it may be inappropriate to say that a builder "kills" a bird if the bird flies into whatever he builds. But on one point, the environmentalists have a case. Tax subsidies make them (and us all) unwilling participants in the whole exercise. Many people would support wind projects if they could compete with other energy sources without subsidies, but they cannot; the expiration of the Renewable Energy tax credit at the end of 2003 "caused a dramatic slowdown in wind projects around the country," according to one report. The wind business went back to work as soon as a one-year extension was signed into law in October 2004.

There are still true believers out there who worry that the problems with renewables are dividing Greens into "bitter factions." But the slaughter of bats and eagles, and the difficulty of making anything more than a marginal contribution to energy needs, came as a blow to many supporters of renewable energy.[20]

But by then another issue was looming even larger—the global-warming scare. It was contrived by the environmentalists, but ironically, it has played into the hands of nuclear power. Global warming, so the theory runs, is caused not by the vagaries of the sun but by the release of carbon dioxide into the atmosphere. But let us not second-guess the ins-and-outs of Green phobias and fashions. The point is that some of those with the media's ear decided that when carbon dioxide rises into the upper atmosphere it is a hazard to the planet. This is where nuclear comes in: carbon dioxide is emitted by the combustion of fossil fuels, but not by the fission of uranium. So some Green leaders began to see nuclear energy as a solution rather than a problem.

Whole Earth Catalog founder Stewart Brand noted the great advantages of nuclear energy: it is atmospherically clean and technologically mature, "with a half century of experience and ever improved engineering behind it." Green aversion to it is "quasi-religious." British environmentalist and Gaia theorist James Lovelock said that opposition was based on "irrational fears fed by Hollywood-style fiction, the Green lobbies, and the media." Nuclear has proved to be "the safest of all energy sources," he added.[21]

Perhaps most surprisingly, a bishop of the Church of England, the late Hugh Montefiore of Birmingham, a member of the board of Friends of the Earth, said that the solution to global warming "is to make more use of nuclear energy." He was kicked off the board right away, but he didn't mind. "The future of the planet is more important than membership of Friends of the Earth." Greenpeace co-founder Patrick Moore also left his organization after embracing atomic energy.

More and more public figures have begun calling for the revival of nuclear power. Senator John McCain of Arizona now supports it, and that trusty weathervane can surely sense which way the winds of opinion are blowing. There are other reasons, including the rising price of oil and natural gas and the instability of the Middle East. But above all we have global warming to thank. For the environmentalists, one contrived scare had trumped another. One never thought anything good would come of global warming, but maybe it has.

Chapter 3

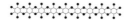

GOOD VIBES
THE VIRTUE OF RADIATION

One does not think of science as being subject to fashion. But it is, and nothing could be less fashionable than hormesis. This phenomenon was identified over a hundred years ago, and then was slowly forgotten. It has been so widely observed that it deserves to be called a law of nature. But it rubs environmentalists the wrong way, and few people have heard of it. In outline, hormesis is simple: things that are toxic in large doses are beneficial in small doses. And that seems to be true across the board—from alcohol to dioxin to mercury to nuclear radiation.

The public policy implications are huge. If hormesis is ever widely accepted, perhaps half the employees at the Environmental Protection Agency (EPA) would be out of work. On a "politically incorrect" scale of one to ten, hormesis is a ten.

Hormesis contradicts a reigning assumption of public health and public policy. It is both dogma and law that something toxic in large doses will continue to be toxic in smaller and smaller doses. The relationship between dose and response is said to be linear. The EPA has further decreed that if the substance is a carcinogen, there is no "threshold." No dose, however low, is considered to be safe.

This "straight-line extrapolation to zero," as it is sometimes called, has kept the EPA busy for thirty-five years. It may be represented by a simple graph.

Guess what?

.ꞏ. Seven hundred thousand shipyard workers servicing nuclear reactors on Navy vessels had a cancer rate 25 percent lower than workers with no work exposure.

.ꞏ. Studies have shown that survivors of Hiroshima and Nagasaki are living longer than Japanese who were nowhere near the blasts in 1945.

.ꞏ. The EPA spent millions on radon abatement, but it turned out that more radon equaled fewer instances of lung cancer.

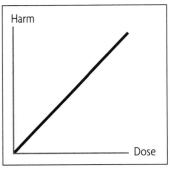

linear, no-threshold

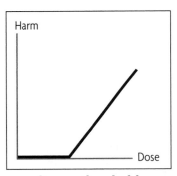

linear, threshold

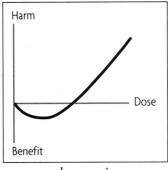

hormesis

There's an alternative "linear" theory, with a threshold. Below a certain dose, the substance is considered to have no effect. This is accepted for many chemicals, as long as they are not cancer-causing. The threshold theory is represented by an equally simple graph, looking like a hockey stick.

Then there's hormesis. Below a certain dose, the response moves into a different territory. The effect is beneficial. At zero dose, obviously, there is zero effect. So it curves back up again to the baseline at zero. The area under the curve shows the "hormetic zone," within which the substance stimulates rather than inhibits.

There is one big difference between this hormesis curve and the two above it. The hormesis data points are *experimentally observed*. Its proponents claim that it has been demonstrated for all substances tested. In the linear no-threshold graph, by contrast, the lower part of the line is simply an extrapolation. Measurements have either not been made at those low levels, or they have been ignored. The *facts* of hormesis, therefore, must be contrasted with the *philosophy* of science practiced by the U.S. government.

The *New York Times* has hardly ever mentioned hormesis—one reason why few people have heard of it. In 2002, reporter Matthew Wald, who has often written about nuclear issues, entered "hormesis" as the search term on his office computer to hunt through the *Times*'s electronic files. He found only one mention of hormesis—in an article published twenty years earlier.

It's worth looking back to 1982 to see what that article said. It was a discussion among three scientists, one of whom was John Gofman, a professor emeritus of medical physics at the University of California–

Berkeley. Gofman had previously served as associate director of the Lawrence Livermore National Laboratory. He had worked on the Manhattan Project. By 1982, he was perhaps the most vehement critic of nuclear power in the country. Licensing a nuclear plant, he once said, "is in my view licensing random premeditated murder."[1] Another participant was Edward Webster, the chief of radiological sciences at Massachusetts General Hospital. The following exchange was printed:

> *Dr. Gofman:* I do not believe that there is any [radiation] dose at this time that has been shown to be without effect. Moreover, I think it is public health irresponsibility to assume that such a dose exists when one is not absolutely certain.

> *Dr. Webster:* I would disagree with that. I would say that there is considerable uncertainty. There are some interesting examples of increased life-span at very low doses in animal populations. In fact, there's a whole body of thought which is developing called hormesis, which means that maybe a little bit of radiation could be beneficial. I'm not saying I agree with this, but it is a school of thought which is the exact antithesis of saying that all low doses of radiation are bad. I don't think we can resolve this controversy, because the low-dose information we have, in general, is too poor to make a decision. I suspect some of the studies Dr. Gofman is talking about are exactly in this category.

> *Dr. Gofman:* Not at all, Dr. Webster. The studies I'm referring to which show increased effects at low dose are based upon Hiroshima-Nagasaki. . . . I think it is public health responsibility to assume effects all the way down.[2]

Almost a decade later, in November 2001, the *New York Times* published a few more tidbits about hormesis in an article about post–September 11 concerns. (The word *hormesis* did not actually appear.) Gina Kolata wrote:

> Some scientists even say low radiation doses may be beneficial. They theorize that these doses protect against cancer by activating cells' natural defense mechanisms. As evidence they cite studies like one in Canada of tuberculosis patients who had multiple chest X-rays and one of nuclear workers in the United States. The tuberculosis patients, some analyses said, had fewer cases of breast cancer than would be expected and the nuclear workers had a lower mortality than would be expected.[3]

As to the atom-bomb survivors, Kolata said that "it has been hard to find excess cancers" among them. Dade Moeller, an emeritus Harvard professor and radiation expert, said that, although the survivors were exposed in 1945, "nearly half" were still alive. Of the children who survived the bombs, over 90 percent were still living fifty years later. There has been no statistically significant increase in birth defects among the children of survivors, nor any increased risk of death among these children up to the age of twenty, the period when hereditary diseases are likely to be seen. Excess birth defects have not been seen in the grandchildren either.

In 1997, Joby Warrick of the *Washington Post* wrote an article about low-level radiation. It was something to be clipped, enclosed in clear plastic, and filed away for safe keeping. It began:

> The statistics seem clear and compelling, and completely at odds with common sense: in Japan, site of the world's only nuclear attacks, radiation victims are *outliving* their peers. It's one of the stranger twists in fifty years of scientific monitoring of atom bomb survivors. As expected, the people closest to

ground zero have died in high numbers of cancers that began in a white hot flash of nuclear radiation. But as one moves further from the blast site, the death rate plunges until it actually dips below the baseline.[4]

Nonetheless, U.S. policy with respect to nuclear radiation is based on the "linear no-threshold" assumption: there is no safe level of radiation. It may have been one of the biggest science policy mistakes of the post–World War II era. The scientist principally responsible for convincing government officials that there should be "no threshold" for radiation was Linus Pauling. It happened this way.

In March 1954, the chairman of the Atomic Energy Commission, Lewis Strauss, said that the first hydrogen bomb, tested about eighteen months earlier at the Eniwetok Atoll in the South Pacific, had resulted in a small increase in radiation in parts of the U.S. mainland. The explosion had sent a cloud of radioactive dust 135,000 feet into the stratosphere. It dispersed and gradually fell back to earth over a two-year period. *Fallout*, it was called. The word was on everyone's lips and became a byword for unpleasant side effects. Strauss added (correctly) that the increase in background radiation was "far below the level which could be harmful in any way to human beings." But that was already too late. In the popular imagination, fallout soon became a fearful thing. Strauss argued for a "threshold theory." Below a certain level, radiation did no harm. That was generally accepted for a while. Shoe stores, for example, used X-ray boxes so that customers could see the outline of their bones when trying on shoes.

Then, in 1955, a geneticist at Caltech, Edward Lewis, wrote a memo on fallout and circulated it among the faculty. When data on the leukemia rates of the Hiroshima and Nagasaki survivors became available, Lewis said, it would be possible to make an early estimate "of the direct effects of radiation." A short time later, the *New York Times* reported that the

Atomic Bomb Casualty Commission had discovered increased incidences of leukemia and cataracts among 30,000 Japanese survivors. And from his hospital in French Equatorial Africa, Nobel Peace Prize winner Dr. Albert Schweitzer called atomic fallout "the greatest and most terrible danger" for mankind.

Linus Pauling, who won the Nobel Prize for chemistry in 1954, called for a suspension of atomic tests. In a speech in Chicago he estimated that a thousand people would die of leukemia from the upcoming British test of the hydrogen bomb. The test went ahead anyway, and Pauling, supported by Barry Commoner and others, started a petition drive. The petition was delivered to the White House with the signatures of nine thousand scientists.

Edward Lewis then wrote a leading article for *Science* in 1957, about "Leukemia and Ionizing* Radiation."[5] A panicked President Eisenhower approved a temporary test-ban, and the subject was popularized in a televised debate between Pauling and Edward Teller, the "father of the H-bomb." Invited to the White House by President Kennedy, Pauling advertised his cause ahead of time with a well-planned demonstration before being ushered into the Oval Office.

Pauling was rewarded with the 1962 Nobel Peace Prize; for good measure, he also won the Gandhi Peace Prize and the Lenin Peace Prize.

Using information provided by Lewis, Pauling argued as follows: if just one stray neutron can initiate cancer, then, by adding up all the people in the world and multiplying by some risk factor, it could be deduced that halting nuclear tests would save tens of thousands of lives. In this way,

---

* Radiation comes in two varieties. The *ionizing* kind is what most people think of when they hear the word radiation. It means that the ray or particle has enough energy to knock an electron out of its orbit. This creates an ionized atom or molecule, which in its altered state has a greater affinity for chemical recombinations. *Non-ionizing* radiation, emitted, for example, by cell phones and power lines, is not dangerous, but some people still want to convince us that it is.

the "linear no-threshold theory" was adopted for nuclear radiation. At Harvard, another Nobel Prize winner, George Wald, summed it up for us: "Every dose is an overdose."

This case was later made by John Gofman, who earlier had been fervently pro-nuclear. During the 1950s, he had supported a "plutonium economy" based on breeder reactors and had urged the licensing of a thousand nuclear power plants. In the late 1960s, however, he made a U-turn. In a "letter of concern" published in 1999, he declared that "there is no safe [radiation] dose, which means that just one decaying radioactive atom can produce permanent mutation in a cell's genetic molecules."

H. Wade Patterson, who was editor in chief of the journal *Health Physics*, worked in Berkeley's radiation lab after 1945. He recalled:

> Citizens there and elsewhere had no fear of radiation until the great debate on nuclear testing. The public exchanges between Pauling and Teller exemplified this controversy. It's an interesting footnote that during the Pauling-Teller debates, John Gofman made many public speeches arguing against the linear no-threshold theory. It was only later that, for mysterious reasons, he made a diametrically opposite change. Anti-nuclear activists and the media seized on the no-threshold theory as the basis for their opposition to testing [and] used it as the basis for all their dire predictions about nuclear power.

As the decades passed, agencies of the U.S. government and their allies in the safety and risk abatement business became increasingly captured by exponents of the linear-no-threshold theory. Government money was doled out to exposed victim groups, including uranium miners and shipyard workers. The measurement and abatement of radon gas became a $100 million-a-year industry, and safety experts in many fields acquired lifetime tenure.

Nuclear power plants emit levels of radiation that are far below the background levels of radiation that we experience daily from cosmic rays and from the earth itself. Nonetheless, the idea took hold that even tiny additional amounts of radiation pose a threat. Nuclear power was put on hold, and we have become more and more dependent on coal for the generation of electricity. And coal is a far greater environmental hazard than radiation.

Background radiation

Theodore Rockwell worked on the Manhattan Project at Oak Ridge, producing enriched uranium, and in 1949 was hired by Hyman Rickover to work in the Naval Nuclear Propulsion Program. He became its technical director and wrote *The Reactor Shielding Design Manual*, which is still in use. Later, he helped found an organization called Radiation, Science and Health, Inc.

Two hundred and twenty nuclear power plants have been installed on ships, compared with 103 civilian plants in the United States. American nuclear-powered ships have been running since 1955 without any significant release of radiation. "Sailors sleep within a few feet of the reactor," Rockwell said. On submarines, "you get less radiation than you do at home because the surrounding seawater protects you from cosmic radiation."[6]

Background radiation is instructive, because every day we are exposed to radiation levels far higher than anything the EPA allows from manmade sources. Natural sources include cosmic rays; radiation from uranium and other radioactive rocks; from radon, a gas emitted by radium; from medical equipment; and from our own bodies as a result of normal metabolism.

Radiation is always and everywhere fading away and dying down, at a rate determined by the half-life of the radioactive material in question.

This may vary from billions of years (in the case of uranium) to a fraction of a second (in the case of certain gases). If we go back millions of years, when man is said to have evolved on the African savannas, radiation levels were higher, so an evolutionary case can be made that we now live in an environment that is deficient in radiation. Overall, natural radiation is decaying more rapidly than made-man radiation is accumulating. So the idea that we are not getting enough radiation to keep us in good health is plausible.

One remarkable study intended to demonstrate the danger of nuclear radiation to workers instead showed the opposite. Released by the Department of Energy in 1991, the study summarized ten years of epidemiological research by the Johns Hopkins School of Public Health. It covered no fewer than 700,000 shipyard workers, including 108,000 who had been occupationally exposed to radiation while installing or servicing nuclear reactors in Navy vessels. (Although no civilian nuclear power plant has been ordered since 1973, nuclear-powered ships are still being built.)

It's All in Your Head

"The psychosomatic disorders observed in the 15 million people in Belarus, Ukraine, and Russia who were affected by the April 1986 Chernobyl accident are probably the accident's most important effect on public health. These disorders could not be attributed to the ionizing radiation, but were assumed to be linked to the popular belief that any amount of man-made radiation—even minuscule, close to zero doses—can cause harm, an assumption that gained wide currency when it was accepted in the 1950s, arbitrarily, as the basis for regulations on radiation and nuclear safety."

Zbigniew Jaworowski, "Radiation Risk and Ethics,"
Physics Today, September 1999

Researchers carefully matched exposed workers with similar workers in the same shipyards who had not been exposed. The exposed group had received radiation doses about ten times that of the unexposed. But researchers found that the irradiated workers had 24 percent lower death rates and 25 percent lower cancer mortality than the unexposed workers. The report was released, but not in a way that forthrightly drew attention to its remarkable findings. Data was buried in the text and not plotted out in revealing graphs for all to see.[7]

A study of American radiologists published in 1950 had shown that they died of leukemia at a rate far higher than other physicians. But they were in many cases pioneers of the technology, who had earlier been exposed to very high doses, at a time when radiation hazards were not properly understood. But a later British study of radiologists who had registered professionally from 1955 to 1979 found that they were half as likely to die of cancer as other physicians who had not worked with X-ray machines. Again, a low radiation dose appeared to have beneficial effects.[8]

Chernobyl

In April 1986, reactor number four in Chernobyl exploded, sending chunks of the reactor core into the surrounding fields and radioactive clouds into the air. The fire burned for ten days, and the wind carried radioactive particles to large swaths of Ukraine. It was the "scene of the worst civilian disaster of the nuclear age," as the *New York Times* put it. In June 2000, the paper reported on the front page that the accident had "spewed radiation over vast stretches of northeastern Europe and caused untold thousands of deaths and illnesses." *Untold* was the operative word.[9]

But when the UN issued its 600-page report, "Chernobyl's Legacy," finding that "only fifty deaths" could be directly attributed to acute radiation exposure, the *Times* buried its report, demurely headlined "Experts

Find Reduced Effects of Chernobyl," on an inside page. Some seven million people in what are now Russia, Ukraine, and Belarus "still receive some kind of Chernobyl benefits, from monthly stipends to university entrance preference to therapeutic annual vacations," the paper reported. The number of people in Ukraine designated as permanently disabled had increased to over 91,000 by 2001, despite the ever-declining radiation levels, and the country was spending about 5 percent of its annual budget on Chernobyl victims.[10]

Given the evidence from elsewhere in the world, it is quite likely that those who received low-level doses as a result of the accident actually benefited.

A few years before this report was released, Theodore Rockwell briefly summarized the latest findings from the UN's Scientific Committee on the Effects of Atomic Radiation, a volume the size of a thick phone book. Some thirty people who were inside the plant died, Rockwell said. He went on to say:

The Fallout: A Bad Prediction

Soon after the explosion of a nuclear reactor at Chernobyl, Ukraine, in 1986, it was estimated that radioactive fallout would cause as many as 150,000 deaths. But when a UN report, "Chernobyl's Legacy," was released in September 2005, a panel of over one hundreds experts could confirm no more than fifty deaths. The deaths were all among reactor staff and emergency workers, who were mostly inside the reactor building at the time of the explosion.

Elisabeth Rosenthal, "Experts Find Reduced Effects of Chernobyl," *New York Times,* September 6, 2005

Some died from the original explosion, some from fire, and I don't doubt some died from radiation. But they were all inside the plant. So it was an industrial accident, and we have seen far worse. As to the general public, they checked for iodine in the thyroid, and sure enough they found 1,800 children with thyroid nodules. But that part of the world is iodine-deficient— they were already having a serious public-health problem. Two kids with thyroid nodules were brought in and they died. But it turns out they were nowhere near the radiation. A third child died of something else entirely. As to the 1,800 people, they did not correlate with radiation dose at all. Some high-dose kids had no nodules, some low-dose did have. So it's not at all clear that they ever were related to the radiation, and the chairman of the original UN committee doesn't think they are related.

Rockwell added that by 2001, "the radiation level in Chernobyl is lower than the natural radiation in my sister's backyard in Denver."

The *Sunday Times* (London) published an article suggesting possible benefits from the increased background radiation. Humans had been evacuated but the animals remained, and they were reported to be "thriving." John Smith of Britain's Center for Ecology and Hydrology said that people think of Chernobyl as a "post-apocalyptic wilderness, whereas it appears to be the exact opposite."

Cham Dallas, a toxicologist from the University of Georgia, visited the area many times and studied mice living close to the dead reactor. "You'd expect them to be in a really bad way. What is weird is that they seem to be unscathed," he said. "They just seem to soak it up." He was quoted as saying that life appears to be "far more resilient to high levels of radioactivity than we anticipated."

Plutonium myth

In the annals of media-fright, plutonium preceded dioxin as "the most toxic substance known to man." Ralph Nader once said that a pound of plutonium could cause eight billion cancers. Bernard L. Cohen attempted over the years to rebut this, but garnered little publicity. "The Myth of Plutonium Toxicity" was the title of a 1979 article he published in a refereed journal. Ralph Nader was so upset by it that he asked the Nuclear Regulatory Commission to investigate. "Which they did in considerable depth and detail," Cohen said. When they gave it a clean bill of health, Nader dropped the subject.[11]

Natural radiation (except radon) averages about 80 millirems a year in the U.S., but in Colorado it's about double that, thanks to the presence of uranium and thorium in the Rockies. Also, its greater altitude reduces atmospheric protection from cosmic rays. In the Gulf States, on the other hand, radiation from the earth is considerably lower. This wide range of background radiation is ideal for a natural epidemiological test of hormesis.

"Cancer rates in the Rockies are only about two-thirds of the national average," Cohen said. By contrast, they are high in the Gulf States. The inverse relationship is striking and needs to be explained. Cohen does not attach much significance to the figures because confounding factors could be at work. Ethnic or cultural factors are possibilities. Researchers like Bruce Ames at Berkeley would point to differences in diet, smoking, or other factors. Maybe the people in the West have better air? No one has offered a satisfactory explanation, but radiation hormesis is a possibility.

A Book You're Not Supposed to Read

Bernard L. Cohen, "The Myth of Plutonium Toxicity," *Health Physics*, v. 32, 1977, 359–79, discussed by Cohen in Karl Otto Ott and Bernard I. Spinard, eds., *Nuclear Energy*, New York: Plenum Press, 1985, 355–65. Also see "Radiation, Science, and Health" (http://cnts.wpi.edu/rsh/) [a website dedicated to an objective investigation of low-level radiation].

But Cohen has also done a more striking study, comparing radon levels with the incidence of lung cancer. As the linear no-threshold theory is causing the U.S. to spend tens of billions of dollars to protect against dangers that may not exist, Cohen thought it essential to test the theory more rigorously in humans. This required more subjects than could be obtained from accidental, occupational, or medical exposures.

One source of radiation recommended itself, and that was radon, a radioactive gas emitted by radium. The EPA says that radon is "extremely toxic" and causes 15,000 lung-cancer deaths a year in the U.S.—about

10 percent of the total. About fifteen years ago, the EPA recommended that levels in homes should not be above a certain level (4 picoCuries per liter). Uranium miners surely have been harmed by high radon doses, but what about the risks posed inside homes?

As with background radiation, the natural level of radon varies considerably—by a factor of ten—from one place to another. So comparisons are possible. The incidence of lung cancer in the U.S. is well reported. With his team at Pittsburgh, Cohen compiled hundreds of thousands of radon measurements to give an average level for 1,729 counties in the U.S., covering 90 percent of the population. The data showed a clear tendency for lung cancer rates, whether corrected for smoking or not, to *decrease* with increasing radon exposure. This information was analyzed

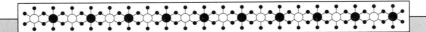

A Cup of Coffee or Plutonium in the Morning

The author of books about nuclear physics and nuclear power, Bernard L. Cohen, an emeritus professor of physics at the University of Pittsburgh, has for years been trying to spread the word about our exaggerated fears of radiation. Plutonium is a *dangerous* substance, in that a critical mass can cause a nuclear explosion. But it is not toxic. Cohen said: "I offered to eat as much plutonium as he [Nader] would eat of caffeine, which my paper shows is comparably dangerous, or given reasonable TV coverage, to personally inhale 1,000 times as much plutonium as he says would be fatal.... My offer was made to all major TV networks but there has never been a reply beyond a request for a copy of my paper. Yet the false statements continue in the news media and surely 95 percent of the public accept them as fact although virtually no one in the radiation health scientific community gives them credence. We have here a complete breakdown in communication between the scientific community and the news media, and an unprecedented display of irresponsibility by the latter."[12]

for over 500 possibly confounding factors, whether socio-economic, geographic, environmental, or ethnic association. Cohen's study strongly supported hormesis.[13]

Cohen found that his own house in Pittsburgh had high radon—five times the level recommended by the EPA. By then he had installed radon-abatement equipment, but once he saw his own results he switched the machine off. He lived in a (relative) radon bath. The results of his county-by-county study were published by *Health Physics* in 1995, but he has found it difficult to publicize them. He has offered a $5,000 reward (which another group will match) to anyone who could offer an explanation for his results consistent with the linear no-threshold theory.[14]

In June 2001, after six years of study, the National Council on Radiation Protection and Measurement recommended that the linear no-threshold theory be retained, but conceded: "It is important to note that the rates of cancer in most populations exposed to low-level radiation have not been found to be detectably increased, and that in most cases the rates have appeared to be decreased."

Those last words put the hormesis argument in a nutshell.

As far as the advocates of hormesis are concerned, a more important question is how long will it be before low-level radiation ceases to be vilified and is used therapeutically?

·•· A recent study involves 1,700 Taiwan apartments constructed with steel girders that were accidentally contaminated with cobalt-60, one of the more dreaded radioactive substances (some may recall the cobalt bomb fears). Over a period of 16 years, some 10,000 occupants were exposed to levels of radiation that should have induced cancers many times in excess of background expectations. Taiwan health statistics predicted 170 cancers among an age-matched population of this size. But only five were observed. Describing this "incredible radiological incident," Y. C. Luan suggested that this might point to "effective immunity from cancer" from the very source thought most likely to give rise to it.[15]

-◆- Ramsar, a town in northern Iran, near the Caspian Sea, has also been studied. Rocks used in local houses contain abundant radium; the two thousand Ramsar inhabitants receive an annual absorbed dose of beta and gamma radiation about fifteen times higher than that permitted for workers at many nuclear power stations. Over several generations, inhabitants also ingested considerable radium in food and inhaled lots of radon. Experimenters tested blood cells in vitro, comparing residents with matched controls from normal-background areas. When these blood samples were subjected to a "challenge dose" of gamma rays, it was found that those from Ramsar had only half the number of chromosomal aberrations that had been induced in the normal controls.[16]

Catching some (gamma) rays

In Boulder, Montana, an engineer discovered in 1949 that a former silver and lead mine was radioactive and leased it. Two years later, a Los Angeles woman visiting with her miner husband noticed that her bursitis disappeared after several visits underground. Today, the mine's owners recommend it for chronic pain and autoimmune disorders, and it is used as a health spa, apparently the only one in the United States. The "therapy area," eighty-five feet below ground, is accessible by wheelchair. Customers pay $5 an hour and about fifty people a day go there in the summer. The radon concentration, 1,700 picoCuries per liter, is over four hundred times the EPA-recommended level.

Among the visitors one day was Klaus Becker, vacationing in the U.S. and catching some "rays" (gamma rays and alpha particles). He was once head of applied dosimetry (the accurate measurement of radiation doses) at Oak Ridge National Laboratory, but now is retired in Germany; also present was Philippe DuPort, who recently founded the Center for Low Dose Research at the University of Ottowa.

Klaus Becker said that the Romans loved to go to radon spas—there was one on the island of Ischia, near Capri—long before anyone knew of it. In fact, many of the famous European spas correspond to radon sites. Bad Gastein in southern Saxony is one. There, Becker said, customers pay $550 for ten hours' inhalation of radon at over a thousand times the EPA-recommended level. There are eleven radon spas in Germany, three in Austria, and three in the Czech Republic. Japan also has them.[17]

"The funny thing is the German government is spending $2 billion in remediation measures, while at the same time, in the same area, a new radon spa has just been officially opened, and the public health service pays for treatment," Becker said. About 75,000 people a year seek such treatment in Germany, mostly for rheumatic and asthmatic conditions. Philippe DuPort said that in Russia they treat about a million patients a year. "But they don't send them down the mines," he said. "They have radiation sources in hospitals." In Japan, also, therapeutic radiation is now officially used.

Chapter 4

"GOOD CHEMISTRY"

A tiny discharge of a substance declared hazardous can cause a major HAZMAT response. Schools can be closed, buildings evacuated, large stretches of rivers deemed unclean, entire communities evacuated. Consider these events:

- Small quantities of substances called PCBs were found in the Hudson River—General Electric had to spend a quarter of a billion dollars to clean it.
- Dioxin was found in Times Beach, Missouri—the whole town was evacuated.
- Twelve to sixteen droplets of the "hazardous chemical" mercury were found on the basement floor of a Washington, D.C. high school—and the school was closed repeatedly for weeks.

In each of these cases, the substance in question has been shown to have beneficial effects at low levels of exposure. However, in many cases, these low levels can trigger an emergency response. Something is wrong here. Society is going to great trouble and expense to clean up substances that may actually be beneficial at the low-dose levels encountered. Maybe it's time to pay attention to chemical hormesis, even if the EPA doesn't approve. We could save ourselves a lot of time and money.

Guess what?

- Support for the theory of hormesis has grown considerably; in the field of environmental toxicology, it's not even controversial.

- Perhaps without even realizing it, people acknowledge that toxic chemicals can be beneficial if they take a multi-vitamin.

- Substances as toxic as dioxin have demonstrated beneficial effects.

The evidence for chemical hormesis is now so strong that it is hardly disputed. It is downplayed, to be sure, but it is undeniable.

The scientist who has done more than any other to revive interest in hormesis is Edward Calabrese of the University of Massachusetts–Amherst. A professor in the school of public health, Calabrese specializes in chemical toxicity. Hormesis is often divided into two categories: chemical and radiation. Radiation is the more contentious, perhaps because it is associated with mass destruction. Calabrese has been content to remain mostly within the more peaceful chemical domain. All he has to deal with are people who are anxious about mercury, lead, cadmium, dioxin, PCBs . . . did I mention arsenic? It's a tonic in small doses. Just don't take too much. The same goes for scotch on the rocks.

The Comeback Kid: Dr. Edward Calabrese

As a student at Bridgewater State College in the 1960s, Calabrese experimented on peppermint plants with a commercial substance called phosfon, made from phosphorous and chlorine. By design, it inhibits the growth of plants. (Gardeners sometimes want that.) But in one experiment, something went wrong. The plants were stimulated instead, and the professor in charge was perplexed.

It turned out the class had made a mistake—they had over-diluted the phosfon. At that watery level it had the opposite effect. The plants apparently liked it. Calabrese offered to do the experiment again, and his teacher, Kenneth Howe, made him do it repeatedly, with twelve different doses in all. In the end he used about 1,500 peppermint plants—far more than would be required in any toxicity test run by the EPA or the FDA today. And the stimulation kept occurring at the low doses. At high doses, the plants withered away as promised. The hormesis curve was unmistakable.

The research was published in 1976. "We never called it hormesis," Ed Calabrese recalled. "Neither I nor Howe had ever heard the term before." Calabrese later found reports from as early as 1958 showing that low-dose phosfon had stimulated other plants, too.[1]

He was hired by the University of Massachusetts, but it wasn't until almost ten years later, in 1985, that he realized the same phenomenon had been found in other fields. Someone sent him a flyer announcing a conference on "radiation hormesis." He didn't know what that meant, but he could see it was similar to what he had found. He told the conference director, Leonard Sagan, about his own experiment and was asked to deliver a paper on chemical hormesis. *Science* published a rare debate on hormesis in 1989.

The next year, Calabrese brought a dozen scientists to Amherst and the modern study of hormesis was launched. One participant came up with the acronym BELLE—Biological Effects of Low Level Exposure. A website with that acronym is a ready source of information about hormesis (www.belleonline.com). BELLE conferences have since been held annually at Amherst. The U.S. Department of Energy has also researched the effects of low-level exposure to chemicals and radiation for several years.

Support for the theory of hormesis has grown considerably since, and in the field of environmental toxicology, it is not even controversial. Hormesis is widely recognized by those who study low-level effects on bacteria, insects, plants, fish, and invertebrates. In mammals it is more controversial, but support is growing there, too. The radiation debate, on the other hand, has been much more polarized.

In 2003, *Nature* published a commentary by Calabrese, headlined "Toxicology Rethinks Its Central Belief." It concluded that hormesis "represents a paradigm shift in the concept of the dose-response throughout biological science." Maybe something close to a biological law had been discovered? Or rediscovered—for something very similar had been proposed over a hundred years earlier.[2]

The term *hormesis*, derived from the Greek word for "excite," was first used in 1943. Two authors found that extracts of cedar stimulated the growth of fungus at low doses and inhibited it at high doses. The phenomenon had previously been described in nineteenth-century Germany. Although hormesis originally described *stimulation* brought about by low-level exposures, it has more recently been used to describe effects that are "beneficial."[3]

But, of course, stimulation is "beneficial" only if it is what we want, and that is not always the case. The stimulation of fungus may be beneficial from the fungus's point of view, but humans may desire less of it. At times it can be downright dangerous to equate stimulation with benefit. If hormesis holds true for bacteria—and the evidence shows that it does—underdoses of antibiotics can pose a special threat. A weak dose may stimulate bacteria to more prolific growth, as the weak phosfon did for the peppermint. Antibiotics, therefore, are not to be taken

A Little Dab Will Do Ya!

In 2003, *Science* magazine investigated some of Dr. Edward Calabrese's claims about hormesis and published a four-page article, "Sipping from a Poisoned Chalice." (*Science*, October 17, 2003.) The magazine found that "hormesis, a concept once discredited in scientific circles, is making a surprising comeback." The reduction of tumors in rats associated with the administration of low doses of dioxin was shown in a simple graph.

Fortune and *U.S. News & World Report* have also reported favorably on hormesis. *Fortune*'s chart showed that high doses of saccharin gives rats cancer, while low doses reduce their cancer risk. (*Fortune*, "A little poison can be good for you," June 9, 2003; *U.S. News & World Report*, "Is there a tonic in the toxin?" October 18, 2004) lightly; they can pose a real hazard. (Harry Lime, the villain in the post-war movie *The Third Man,* was selling diluted penicillin on the black market.)

So frequently has hormesis been observed that it is difficult to find substances that are *not* stimulative at low doses. It deserves to be called a law. Something called the Arndt-Schulz Law was formulated in Germany over a hundred years ago. Hugo Schultz applied many chemicals to yeast and noted their effects on fermentation. Later, he joined up with physician Rudolph Arndt. The two claimed that their findings applied to all organisms and toxic agents. Support for low-dose stimulation as a biological principle was refined in 1896 by Ferdinand Hueppe, who studied under Nobel Prize winner Robert Koch.

Perhaps without even, realizing it, we acknowledge that toxic chemicals can be beneficial in small doses when we take multi-vitamin pills. Their ingredients are printed on the bottle. One such product lists the following: iodine, phosphorous, magnesium, selenium, zinc, copper, manganese, chromium, molybdenum, potassium, nickel, boron, and vanadium. All of these are toxic substances—at high doses.

It is sometimes said that this argument is invalid because these minerals are "trace elements" essential for life. Nonetheless, most or all of these substances would be fatal if consumed in high doses. Some are highly toxic, such as selenium and manganese. The widespread use of multi-vitamins shows that people assume something other than the "linear no-threshold" theory when they take their supplements.

The real battle will be fought over substances *not* known to be essential for life. Calabrese and Blair reported in the *Journal of Environmental Monitoring* (2004) that hormetic responses "are commonly reported amongst the most toxic of the heavy metals and even in the most sensitive target organs such as the brain." These include aluminum, arsenic, cadmium, chromium, cobalt, copper, gold, lead, mercury, nickel, selenium, tin compounds, vanadium, and zinc. All have shown the same basic hormetic dose response. At low doses, they have a stimulative effect in a wide range of living organisms.

When people hear about hormesis, they often compare it with homeopathy, or at least think the two are the same. Homeopathy originated with a physician in Leipzig at the end of the eighteenth century. Practitioners believe that diseases should be treated "by the administration (usually in very small doses) of drugs which would produce, in a healthy person, symptoms closely resembling those of the disease treated." (Definition from the Oxford English Dictionary.)

The fundamental doctrine of homeopathy is sometimes given as "likes are cured by likes." But notice that the similarity in homeopathy is of symptoms, not agents. For example, a homoeopathist may prescribe a small dose of arsenic for a stomach virus "because arsenic poisoning also causes abdominal pain and vomiting. It's a matter of analogy more than biology."[4]

In addition, doses given according to homeopathy are far more dilute than the dose range in which hormetic effects have been observed—maybe a million times weaker. In fact, homeopathic doses are usually so low that the benefits may be compared to a placebo effect. For that reason, scientists are inclined to dismiss the claims of homeopathy.

The historical connection here is important, because Arndt and Schultz took their findings as confirmation of homeopathy and championed its medical use. Rudolph Arndt was himself a practitioner of the homeopathic arts, which was one of the principal reasons why the idea underlying hormesis fell into disfavor. When the idea was reformulated in the 1940s, it was handicapped because those who could remember hormesis assumed it was the same as homeopathy and believed it had been discredited.

Another reason for the early demise of hormesis was that soon after X-rays were discovered, in 1898, the beneficial effects of radiation were promoted before its dangers were understood. Researchers began publishing reports of radiation hormesis; some doctors touted radioactive patent medicines for various ailments. The reign of radioactive elixirs and radium therapy came to an end on March 31, 1932, when a millionaire industrialist died from bone cancer. His death received front-page coverage in the *New York Times*: "Eben M. Byers Dies of Radium Poisoning!" For several years, Byers had routinely consumed the product Radithor, which contained radium. By then, researchers also knew that radium dial painters (who used their lips to make a pointed tip on their brushes) were developing cancer.

At that point, hormesis began to look more like a blunder than a discovery. Later, environmentalism brought the issue of hazardous substances into public view, engendering tremendous fear. This fear peaked in the early 1980s, when Congress enacted the so-called Superfund legislation. Still on the books, the law imposes a heavy and retroactive burden of fines and mandates on companies deemed responsible for pollution. The U.S. government adopted an expensive and punitive no-threshold standard for most hazardous substances. Tiny traces of chemicals such as dioxin would have to be cleaned up, at vast expense. Sometimes, the evacuation of whole communities was called for.[5]

Hormesis on the Rocks with a Twist

Hormesis verges on conventional wisdom for some substances. *Scientific American*'s February 2003 article "Drink to Your Health?" carried the subhead: "Three decades of research shows that drinking small to moderate amounts of alcohol has cardiovascular benefits." A few days earlier, the *Washington Post* reported on page one "Daily Alcohol Cuts Risk of Heart Attack, Study Finds." As for excessive drinking, we all accept that it can kill you.

The dioxin panic

Since the 1970s, four big cases involving dioxin have generated tremendous panic. Each attracted world-wide attention and stoked public fears.

-•- Dioxin acquired notoriety as a byproduct of Agent Orange, an herbicide made by Dow Chemical and used as a defoliant in the Vietnam War. Between 1967 and 1970, about 1,300 Air Force personnel were involved in spraying. Its use was discontinued in Vietnam in 1970. Many health problems associated with Agent Orange were alleged by military personnel, including claims of cancer and birth defects in offspring. A class action lawsuit, eventually joined by 200,000 veterans, was settled by the manufacturers in May 1984.[6]

-•- In July 1976, a Hoffman-LaRoche chemical plant in Seveso, Italy, exploded, exposing 37,000 people to a cloud of gases, dioxin among them. The following year, a woman in Oregon claimed that local miscarriages were connected to the spraying of the herbicide. The EPA studied the situation and claimed to find a connection. In 1979, the agency suspended the herbicide in the U.S. market.

-•- In the state of New York, Love Canal was exposed as a Hooker Chemical dump site. After twenty years of dumping, seepage occurred; residents began to smell bad odors and before long felt sick. Local stories about Love Canal became national news. "It was the presence of dioxin, more than anything else, that terrified the residents," Michael Fumento wrote in *Science Under Siege*. Townspeople took to the streets carrying signs reading, "Dioxin is here," "Dioxin kills." A television documentary said the chemical was so powerful that "an ounce could wipe out a million people." Ralph Nader and his assistants reported that forty-three million pounds of many different chemicals lay beneath the surface of Love Canal, among them benzene and chloroform. Dioxin in particular was said to evoke "fear and anxiety among scientists and doctors." According to estimates, over a hundred pounds of it lay buried.[7]

·⋅ More dioxin was discovered in Times Beach, Missouri, in 1982. It came from a chemical company that manufactured a bacteriocidal agent used in the treatment of acne and impetigo. Dioxin is its byproduct, and about fifty pounds of the dreaded substance by then had been spread around the state. Around Christmas 1982, the U.S. Centers for Disease Control (CDC) recommended that Times Beach be evacuated. Congress intervened, and Rita Lavelle, an EPA official who had opposed final action until all the evidence was collected, was fired (and later indicted for contempt of Congress). In February 1983, the federal government announced that with some assistance from the state of Missouri, it would buy all the residential and business properties of Times Beach.

Studying dioxin

The health consequences of the factory explosion in Seveso, Italy, were studied minutely. Almost nine hundred acres had been contaminated. Soon after the accident, the average concentration of dioxin was found to be about five hundred parts per billion, or five hundred times higher than the "level of concern" set by the CDC at Times Beach. The CDC estimated that continuous exposure of a million people to one part per billion of dioxin over a lifetime would cause one additional case of cancer.

Built into these CDC and EPA estimates was the assumption of linearity between dose and risk. And that is what the theory of hormesis contests. The assumption has been adopted by all U.S. and UN agencies. A linear relationship would imply that a dioxin concentration of five hundred parts per billion would be associated with five hundred extra cancers in a population of one million.

Newsweek titled its article on the Seveso accident "Our Own Hiroshima." But in the end, investigators were unable to conclude that a single person had died as a result of dioxin exposure. About 180 cases of chloracne were diagnosed, especially among children. Chloracne (a contraction of chlorine and acne) resembles teenage acne and has often been found among chemical industry workers. It was first described in Germany in 1895. Among the Seveso cases, almost all the chloracne symptoms disappeared within five years.

There was such panic among pregnant women exposed to the chemical fallout that ninety of them went to Switzerland to have abortions—still illegal in Italy in 1976. Their fears were groundless. An examination of twenty-five induced abortions found "no evidence of chromosomal aberrations."[8] As to birth defects, the *Journal of the American Medical Association* published a follow-up study in 1988. "No correlation or association between contaminated areas and [fetal] malformations was found." Among the children born during the early months of 1977, who probably had been exposed to some dioxin in the womb, "no infant with a major malformation was identified in the most highly contaminated zone."

Similar benign results were found with Agent Orange. The pilots who had sprayed the herbicide were exposed daily. According to the *New York Times* (August 13, 1984), "In their tour of duty they received a thousand times more than would ground troops by being sprayed directly." Yet a survey of their health "showed no unusual incidence of diseases." After prolonged investigation, the *Washington Post* (January 15, 1985) concluded that "there is no solid evidence to support veterans' claims that Agent Orange is responsible for the ailments that they and their offspring—along with many other people in the general population—have suffered."

The main study of the relationship between dioxin and cancer was done by the Dow Chemical Company. Published in 1978, it is often referred to as the Kociba study, after its principal investigator. Over a two-year period, hundreds of rats were exposed to different levels of dioxin (mixed in with their food). As a result, the EPA confidently proclaimed dioxin to be carcinogenic.

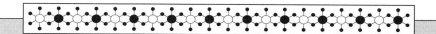

We're the Government and We've Come to Help

In 1991, the EPA undertook a "reassessment" of the health risks posed by dioxin. The *New York Times* commented in an editorial: "Federal officials now believe they may have overreacted in setting extremely low exposure limits for dioxin and in permanently evacuating all the residents of Times Beach, Missouri.... A steady accumulation of data has convinced many health experts that dioxin is only a moderate threat to human beings. It appears far less risky, for example, than asbestos, radon, nickel, coke, chromates, or smoking."

Dr. Vernon Houk, the official from the CDC who had urged the evacuation of Times Beach, said it had been a mistake. What would he do differently, if he could go back? "I would have said we may be wrong," he said. "If we're going to be wrong, we'll be wrong on the side of protecting human health." He added that his decision had been based "on the best scientific information we had at the time. It turns out we were in error."

The *New York Times* also said in the editorial that studies of human exposure in accidents and in industrial settings had shown "either no effect, or a modest increase in cancer among those receiving very high doses."

But there was no mention of what happened at low doses. That is the information that has consistently gone unreported, in the *Times* and elsewhere, not just with regard to dioxin, but with many other agents that are toxic at high doses.

However, the data reveal that the dose-response relationship was actually J-shaped. Rats get cancer anyway without deliberate exposure to toxic substances, so unexposed "control" rats were used to establish the background incidence of tumors. It turned out that this type of rat had quite a high frequency of spontaneous tumors, ranging from about one to three tumors per animal in a lifetime.

Far from being a drawback, this was advantageous for researchers. If the background incidence is low or zero, low doses can hardly give results that go below that level. Mice studies often yield no useful data because the control groups may have only one or two tumors between them.

When dioxin was added to the diet of experimental rats, and when all the tumor sites were aggregated, it was found that the incidence of tumors fell quite sharply for both male and females. Among females, for example, the background rate of 2.67 tumors per rat fell to 1.92 at the lowest dose; and among males the background rate of 1.62 fell to 0.8. That is, at the lowest administered dose, the male rats had only half as many tumors as the controls.

This relationship—a decline in tumors at low doses—also held true when all the tumor sites were examined separately. But there was *one category*, and *only* one—liver tumors in female rats—in which the tumor incidence increased markedly at the two higher doses. But in this category also, the tumor incidence declined (from 10 percent among the controls to 6 percent) in the low-dose group.

The study showed that at low doses, dioxin actually offered rats protection against cancer. In some of the tumor categories, such as the uterus, mammaries, pituitary, and pancreas, the incidence of tumors remained well below that of controls even at the higher doses of dioxin. This suggests that the "lowest observed adverse effect level" of dioxin was not even reached by the experimenters.

Liver tumors did increase seven-fold in the highly dosed female rats, however, and the elevation in this single category allowed the EPA to identify dioxin as a potent carcinogen.

The current approach to animal toxicology "is to assume that an increase in tumors of one organ means that there will be increased frequency of tumors overall, even if there isn't," Ralph Cook has written. "It further assumes that such an increase in benign and malignant tumors in animals will also occur in humans."[9]

A few years later, the National Toxicology Program, a U.S. government agency that is part of the NIH, repeated the Kociba study with a different strain of rat. The same result was found. Low doses of dioxin markedly reduced the incidence of tumors. Three doses were used. In the case of liver tumors in females, the background incidence of tumors among controls was 6.66 percent. But among the rats with a low dose of dioxin, this fell to 2 percent. At the moderate-dose level, the incidence was the same as that of controls, and at high exposure, the incidence rose to four times that of controls.

Looks Are Only Skin Deep

Dioxin was back in the news in 2004. Someone slipped dioxin into the food of Ukrainian opposition leader Viktor Yushchenko just before the country's presidential election. He was reported to have ingested the "most harmful variety," and the level in his blood was six thousand times higher than normal. He developed a bad case of acne, but his internal organs were unaffected. The disfigurement of Yushchenko's face was shown around the world. Those who had spiked his food had evidently miscalculated. Maybe they believed it really was "the most toxic chemical known to man," as has been reported in the United States.

In U.S. government documents, it is difficult to find any reference to this "biphasic" character of the body's response at different doses. ("Biphasic" means that the graph has two phases—"down" at low doses and "up" at high doses.) Instead, the government's Toxic Substances and Disease Registry records that for dioxin, "doses of approximately 0.071 micrograms-per-kilogram per day increased the incidence of neoplastic nodules in the liver and of hepatocellular carcinoma" in female rats.

But the Registry failed to mention that at lower doses, the same substance provided protection at the same site in the body. Most people, of course, are unable to decipher the complex terminology in which toxicologists express doses. Laymen may well assume that low doses were being described in the above example, when in fact much lower doses had also been administered, to the evident advantage of the animal.

Therefore, animal studies with dioxin strongly support the idea of hormesis. At low doses, the dreaded chemical actually seems to provide protection against cancer. So pronounced is the protective effect that one pharmaceutical company has begun to look into the possibility of developing a cancer prophylactic based on dioxin.

A Book You're Not Supposed to Read

Dioxin, Agent Orange: The Facts by Michael Gough; New York: Plenum Press, 1986.

What about the effects of dioxin on humans? Though experimental consumption by humans has not been tried, the explosion in Seveso can be thought of as an inadvertent experiment. In 1984, an International Steering Committee, including medical experts from the U.S and many other countries, reviewed all of the numerous studies and concluded: "Nearly eight years after the accident in Seveso it has become obvious that besides chloracne in a very small group of cases, no adverse health effects related to the chemical produced by the accident has been observed."[10]

There was also a ten-year follow-up. In June 1989, the *American Journal of Epidemiology* included a study by Italian doctors. They found that "overall, the relative risk for malignancies as a whole tended to be below 1.0," meaning that there were fewer malignancies than in the unexposed population.[11]

The exposed human population in the Seveso region was divided into three groups—Zone A, with the highest exposure; Zone B, intermediate; and Zone R, the lowest contamination. There were 26,227 people in Zone

R, and their ten-year mortality was compared with 167,000 people in a regional comparison group. In this low-contamination zone, liver cancer deaths among both males and females were less than half of those reported in the unexposed comparison group. Deaths from lung and breast cancer were also reduced, although less dramatically. Ralph Cook summarized the finding this way: "If dioxin is an initiator, too short a time may have passed to get meaningful results. On the other hand, the results are consistent with dioxin being either preventive or therapeutic for breast cancer, and possibly other tumors."[12]

Chapter 5

THE DDT BAN

Dichloro-diphenyl-trichloroethane was first synthesized in 1874 by a German chemist, then independently rediscovered by Dr. Paul Mueller in 1939. Mueller found that DDT quickly killed flies, lice, fleas, and mosquitoes—carriers of serious infectious diseases such as the bubonic plague, typhus, yellow fever, encephalitis, and malaria. The insecticide was patented in the United States in 1943, and its beneficial uses were soon confirmed by the U.S. Department of Agriculture.

Toward the end of World War II, the U.S. Army issued small tin containers of 10 percent DDT dust to its soldiers around the world. It was effective in killing body, head, and crab lice. Five hundred gallons of DDT were delivered to Italy to combat a rapidly spreading epidemic of louse-borne typhus. Paul Mueller won the Nobel Prize for medicine in 1948 for his work on DDT. It was one of the greatest single improvements to public health since the introduction of clean drinking water.[2]

After 1945, the U.S. Public Health Service assumed responsibility for administering malaria eradication programs in eighteen countries, with the support of the State Department. When a program was introduced in Greece, for example, the annual number of malaria cases dropped from two million to about 50,000 within three years.

Guess what?

.ஃ. DDT was banned by the EPA in 1972—malaria now kills more than a million people per year in Africa alone.

.ஃ. The editorial pages of the *New York Times* and the *Wall Street Journal* rarely agree, but they both support the idea that the insecticide DDT should be brought back.[1]

.ஃ. Even Ralph Nader supports the careful use of DDT, and rock stars are beginning to participate in a program called Roll Back Malaria.

By 1955, however, 10 percent of the world's population was still suffering malarial attacks, and it was estimated that every ten seconds someone died of the disease. Malaria was contracted by 300 to 400 million persons every year, killing three or four million of them. India alone was losing 800,000 people a year. The World Health Organization (WHO) then announced a world-wide war on malaria in 1955, and the U.S. Congress adopted the same policy in 1957.

As a result of the campaign, malaria was eradicated from all developed countries by 1967. Large areas of tropical Asia and Latin America were also freed from the scourge. In Ceylon, later called Sri Lanka, 2.8 million cases of malaria a year fell to seventeen. In India, the number of deaths became inconspicuous. But the eradication campaign was launched in only three countries of tropical Africa, since it was not considered feasible in poorer areas of the world.

Silent but deadly: Rachel Carson

The tide turned against DDT with the publication of Rachel Carson's book *Silent Spring*. Practically the founding document of the modern environmental movement, it was serialized in the *New Yorker* and published as a book in 1962. It introduced Americans to the concept of ecology. Poison one creature, and others that feed on it will be poisoned in turn—kill mosquitoes intentionally and eagles will die by mistake. If one creature is eliminated, others will lose their natural enemies. Everything is interconnected. Because nature is so complex, humans don't know what they are tampering with and so are well advised to leave it alone.

Rachel Carson, who worked for the U.S. Fish and Wildlife Service for much of her career, had long been preoccupied with DDT. Her book started life as a report on a lawsuit brought against federal and state governments for the aerial spraying of DDT on various properties in Long

Island, New York. The suit was dismissed on a technicality but the book lived on. Carson herself died of cancer in 1964, at the age of fifty-six.

Her book was a turning point. For the first time, Americans questioned the reassurances routinely offered by scientists and officials. But alarmism, and then politics, soon filled the vacuum of authority.[3]

Politicized science

In 1971, the EPA carried out months of hearings to determine the risks and benefits of DDT. Nothing caused more concern than the rumor that it might cause cancer. DDT had been freely sprayed from the air, drifting in the wind while children played. The hearing examiner finally concluded that DDT was not a carcinogenic hazard to man. But in 1972, the EPA administrator, William Ruckelshaus, banned the substance anyway on the grounds that DDT "posed a carcinogenic risk."[4]

Ruckelshaus cited experiments showing that DDT caused cancer in mice. But the evidence was murky, with dosages up to a thousand times higher than anything found in the human diet. In effect, researchers upped the dosages on the mice until they got the results they wanted. Some experiments had found no additional tumors in the animals. At least one

Before It Was Politically Incorrect, They Said:

A Committee of the National Academy of Sciences said in 1970: "To only a few chemicals does man owe as great a debt as to DDT... In little more than two decades, DDT has prevented 500 million human deaths, due to malaria, that otherwise would have been inevitable."

"The Life Sciences," National Academy of Sciences, NAS Press, Washington D.C., 1970

DDT and Birth Control

"My own doubts came when DDT was introduced for civilian use. In Guyana, within two years, it had almost eliminated malaria, but at the same time the birth rate had doubled. So my chief quarrel with DDT in hindsight is that it greatly added to the population problem."

Alexander King, co-founder of the Club of Rome, *The Discipline of Curiosity*; Burlington, MA: Elsevier, 1990, 43.

experiment, with rats, showed DDT to be an anti-carcinogen. "Pretreatment of female Sprague-Dawley rats" with DDT "significantly reduces their subsequent liability to mammary tumor induction," it was reported. (Maybe one day the pharmaceutical companies will look into DDT's potential as an anti-cancer drug.)

As to the cancer risk, the extrapolation from the effects of high doses in mice to low doses in humans was not warranted. Many experimental substances, administered in high doses in animals, have been shown to cause cancer—salt, for example, and vitamin A. UC–Berkeley biochemist Bruce Ames has pointed out that rodent carcinogens are present in almost all fruits and vegetables, including apples, bananas, broccoli, Brussels sprouts, cabbage, mushrooms, and oranges. That doesn't mean they should be banned.

It was also claimed that DDT concentrations were cumulative: the chemical builds up in the environment without degrading, persists for decades in the oceans, and so on. One scientist made this claim in an article published in *Science* in 1964, not long after *Silent Spring* was published: after forests were sprayed in a certain location, the concentration of DDT in the soil built up to higher levels each year. Then it was revealed that the sampling site was beside the local forest airstrip and was heavily dosed with DDT by aircraft during the testing and calibration of spray equipment.

Shortly before he testified at the EPA hearings, the same scientist backtracked. He and his co-authors reported in *Science* that less than one-thirtieth of one year's production of DDT could still be found later, in a much more widespread search. "Most of the DDT produced has either been degraded to innocuousness or sequestered in places where it is not freely available to the biota," they concluded.

According to tests conducted by Dr. Philip Butler, director of one of the Fish and Wildlife Service's research laboratories, "92 percent of DDT and its metabolites disappear" from the environment after thirty-eight days. This data was presented at the EPA hearings. Even earlier, the director of the WHO had said that no symptoms from DDT "have been observed among the 130,000 spraymen" or the millions of inhabitants of sprayed houses.

The WHO, therefore, had "no grounds to abandon this chemical which has saved millions of lives, the discontinuation of which would result in thousands of human deaths and millions of illnesses. It has served at least two billion people in the world without costing a single human life by poisoning from DDT," the WHO director added. Its discontinuation "would be a disaster to world health." That was in 1969.

Walking on eggshells

One of the best publicized and most harmful claims about DDT was that it caused the eggshells of birds to be too thin. Birds returning to their nests would crush their own eggs! One researcher, J. Bitman, had demonstrated the thinning by feeding test birds some DDT but had also reduced the calcium level in the birds' diet. This did produce thinner shells. Bitman then redid the experiment, retaining the DDT but restoring the missing calcium. This time, the eggshells were of normal thickness.

"Unfortunately," J. Gordon Edwards wrote in a review of the general malfeasance surrounding DDT, "the editor of *Science* refused to publish the results of that later research. Editor Philip Abelson had already told Dr. Thomas Jukes of the University of California in Berkeley that *Science* would never publish anything that was not antagonistic toward DDT. Bitman therefore had to publish the results of his legitimate feeding experiments in an obscure specialty journal, and many readers of *Science* continued to believe that DDT could cause birds to lay thin-shelled eggs."[5]

The second article, setting the record straight, was published in *Poultry Science*.

Edwards, a professor of entomology at San Jose State University, was for many years a leading critic of the DDT ban. He testified in favor of the pesticide at the EPA hearings, and demonstrated his conviction that DDT was harmless to humans by consuming a spoonful of it in front of students at the beginning of the academic year. He remained in good health, climbed mountains, and died in 2004 at the age of eighty-five.

EPA ban

In his conclusion, EPA hearing examiner Edmund Sweeney wrote: "DDT is not a carcinogenic, mutagenic or teratogenic hazard to man. The uses under regulations involved here do not have a deleterious effect on freshwater fish, estuarine organisms, wild birds or other wildlife...The evidence in this proceeding supports the conclusion that there is a present need for essential uses of DDT."

The decision was overruled by EPA adminstrator Ruckelshaus, who was intent on banning DDT. As an assistant attorney general, he had stated in 1970 that "DDT has an exemplary record of safe use, does not cause a toxic response in man or other animals, and is not harmful. Carcinogenic claims regarding DDT are unproven speculation." Later, however, in a 1971 address to the Audubon Society, he said: "As a member of the Society, myself, I was highly suspicious of this compound, to put it mildly. But I was compelled by the facts to temper my emotions... because the best scientific evidence available did not warrant such a precipitate action. However, we in the EPA have streamlined our administrative procedures so we can now suspend registration of DDT and the other persistent pesticides at any time during the period of review."

In view of what would happen later on the African continent and elsewhere, it is significant that Ruckelshaus did not file an environmental impact statement on the anticipated effects of his ban. Such was the hysteria about pesticides and the supposed decline of the robin in our "silent spring" that disease and death in the Third World was not on anyone's mind.

Africa pays

In a detailed 2004 article in the *New York Times Magazine*, "What the World Needs Now Is DDT," Tina Rosenberg, an editorial writer for the newspaper, said that while re-reading *Silent Spring*, she was struck by something she had not noticed before. "In her 297 pages, Rachel Carson never mentioned the fact that by the time she was writing, DDT was responsible for saving tens of millions of lives, perhaps hundreds of

What Do Africa, Shakespeare, and Oliver Cromwell Have in Common?

Many people think of malaria as a "tropical disease," but that is a testament to its successful eradication in the developed world—and to the efficacy of DDT. In the United States, there had been six or seven million cases of malaria every year in the 1930s. The disease was found as far north as Montreal. In England, Oliver Cromwell died of malaria, and Shakespeare referred to it (as "the ague") in several plays.

millions." Rosenberg added this remarkable indictment of the book: "DDT killed bald eagles because of its persistence in the environment. *Silent Spring* is now killing African children because of its persistence in the public mind. Public opinion is so firm on DDT that even officials who know it can be employed safely dare not recommend its use."

What seems so unfair—perhaps the Congressional Black Caucus should take an interest—is that DDT was used to eradicate malaria in developed countries and then banned before it could do the same thing in undeveloped countries. Africans became the victims; of lesser disasters are conspiracy theories born. In November 2004, Roy Innis, the national chairman of the Congress of Racial Equality, sent a letter to President Bush. It included the following summary of the facts:

> The United States and Europe eradicated malaria after World War II, using pesticides and other measures. But today, this vicious killer still infects 300,000,000 people every year in developing countries—more than live in the entire United States. It kills as many as 2,000,000 every year—the population of Houston, Texas: another father, mother, or child every 15 seconds. Nearly 90 percent of these victims are in sub-Saharan Africa, and the vast majority are children and pregnant women. Since 1972, at least 50 million people have died from malaria. Heaven alone knows how many might have lived, if their countries had been able to control this mosquito-borne disease."[6]

A key point is this: the EPA ban applied only to the United States. Why should it affect countries in the Third World? Tina Rosenberg's article provided this missing piece to the puzzle:

> Various factors, chiefly the persistence of DDT's toxic image in the West and the disproportionate weight that American deci-

sions carry worldwide, have conspired to make it essentially unavailable to most malarial nations. With the exception of South Africa and a few others, African countries depend heavily on donors to pay for malaria control. . . . Major donors, including the United States Agency for International Development, have not financed any use of DDT, and global health institutions like WHO and its malaria program, actively discourage countries from using it.

Here's a case in which America's influence in the world has been downright (and literally) unhealthy. It is also an object lesson in the pitfalls of foreign aid and why nations should avoid dependence on it. They import not just the donors' money but also their prejudices.

Anne Peterson, the assistant administrator for global health at USAID, said that paying for DDT abroad raises the issue of a double standard. "What would Africans think if we're going to do to them what we wouldn't do to our own people?" Of course, the double standard was already in plain view. Americans had used the chemical to rid themselves of malaria and then demonized it for everyone else.

Ruckelshaus, now retired, said recently that "it's not up to us to balance risks and benefits for other people." If he were a decision maker in Sri Lanka, where the benefits of DDT outweigh the risks, he said, he would not have banned DDT. "There's arrogance in the idea that everybody is going to do what we do. We're not making these decisions for the rest of the world, are we?"

Alas, yes, we are.

The learning curve

Foreign countries are slowly learning that they can't afford to put themselves at the mercy of American politicians and environmentalists. And where they can afford to escape this influence, they are doing so. Both India and China now manufacture and use DDT. The evidence that this is the right course of action grows with each and every year. The *Wall Street Journal* reported in late 2004: "When South Africa stopped using DDT in 1996 at the urging of environmentalists, malaria cases rose from 6,000 in 1995 to 60,000 in 2000. DDT use resumed in the country's worst-hit province, KwaZulu Natal, and malaria cases fell by nearly 80 percent by 2001. Zambia, one of Africa's poorest countries, also saw a tremendous drop in malaria cases when insecticide spraying was reintroduced four years ago."

New York Times columnist Nicholas D. Kristof elaborated on the same theme in a January 8, 2005, column titled "It's Time to Spray DDT":

> The poor countries that were able to keep malaria in check tend to be the same few that continued to use DDT, like Ecuador. Similarly, in Mexico, malaria rose and fell with the use of DDT. [But Mexico was told to give up DDT as part of the North American Free Trade Agreement.] South Africa brought DDT back in 2000, after a switch to other pesticides had led to a surge in malaria, and now the disease is under control again. The evidence is overwhelming: DDT saves lives.

The attitude of the international aid establishment is now schizophrenic toward the ban. Today, some groups put up little resistance to DDT. Kristof called a couple of environmental groups for comment, thinking he "would get a fight." But the man from the World Wildlife Fund, Richard Liroff, acquiesced. "South Africa was right to use DDT," he said. "If the alternatives to DDT aren't working, as they weren't in South Africa, geez, you've got to use it." And at Greenpeace, Rick Hind said: "If there's nothing else and it's going to save lives, we're all for it. Nobody's dogmatic about it."

In fact, it's getting harder to find people who still favor the DDT ban. Vice Admiral Dr. Harold M. Koenig, a former surgeon general of the Navy, said this about the politics of the ban: "Poor public policies [prohibiting the use of DDT] are being implemented because it is easier for politicians to go along with the noise coming from the hysterics rather than to learn the whole story and educate the general electorate that there are ways agents like DDT can be used safely."[7]

The ban has turned out to be "a colossal tragedy," says Donald Roberts, a professor at the Uniformed Services University of the Health Sciences in Bethesda, Maryland. "It's embroiled in environmental politics and incompetent bureaucracies." He told Front-PageMag.com: "DDT is the best insecticide we have today for controlling malaria. DDT is long acting, the alternatives are not. DDT is cheap, the alternatives are not. End of story."

Supporters of the ban give reasons that are either silly or tragic. One who continues to hold the line against DDT is Paul Ehrlich, the Stanford University biologist who has been on the wrong side of just about every issue he has embraced, most notably his fantasies of overpopulation and the death of millions by starvation. Ehrlich wrote a (co-authored) letter to the *New York Times* making the curious point that "DDT's use and management are neither cause nor cure of malaria, despite the cries from those who should know better."

"Neither cause nor cure of malaria"...? Think about that; he advances his cause—the DDT ban—by rebutting a claim that no one has made. DDT doesn't cure malaria; it kills the mosquitoes that cause the malaria. Then there's the Sierra Club's Michael McCloskey, who said in 1971: "The

A Book You're Not Supposed to Read

Junk Science Judo by Steven Milloy; Washington, D.C.: Cato Institute, 2001. Also see Malloy's material on DDT, available at http://www.junkscience.com. Also see "What the World Needs Now Is DDT," by Tina Rosenberg, *New York Times Magazine*, April 11, 2004.

Sierra Club wants a ban on pesticides, even in countries where DDT has kept malaria under control . . . [because by] using DDT we reduce mortality rates in underdeveloped countries without the consideration of how to support the increase in populations."[8]

The truth will out, and people have been charged with genocide for less. But here at least we do see in plain view the shameful issue that has driven a lot of the opposition to DDT.

How to rescind the ban

Here are three possibilities:

- ⋅∔⋅ The question of reviving the spraying of DDT was raised at a conference on tropical diseases sponsored by the Bill and Melinda Gates Foundation. The question was not on the agenda but was raised informally. No one was sympathetic to the idea of bringing back DDT, or even seemed to have given it any thought. But one participant did say this: if DDT were to be brought back, it would first have to be given a new name. It's a thought.[9]

- ⋅∔⋅ When controversy erupted over the appointment of Paul Wolfowitz to the World Bank, the former Defense Department official immediately scheduled a meeting with the Irish rock singer Bono, without whose participation it seems that no foreign aid decision can be made. Bush's treasury secretary, Paul O'Neill, dutifully trotted off with Bono on a tour of inspection of the Third World. Perhaps Bono can be persuaded to alert President Bush to our role in the death of African children?[10]

- ⋅∔⋅ But former surgeon general Harold Koenig pointed to the one development that surely will do the trick. "It is only a matter of time, a short time, before we see these [mosquito-borne] diseases again in the regions between the tropics and the poles," he said.

If malaria returns to the United States, you can bet that DDT will return, too—and quickly.

Chapter 6

BIODIVERSITY AND ENDANGERED SPECIES

dward O. Wilson, Harvard professor, two-time winner of the Pulitzer Prize, world expert on ants, and prodigious salesman of speculative biology, is perhaps the best-known authority in America on the extinction of species. Today the subject is often addressed under the heading of "biodiversity." In his book *The Future of Life*, published in 2002, Wilson wrote:

> How much extinction is occurring today? Researchers generally agree that it is catastrophically high, somewhere between one thousand and ten thousand times the rate before human beings began to exert a significant pressure on the environment. The previous natural, or Edenic, period of biodiversity as measured by paleontologists, started . . . 450 million years ago. It ended fifty to ten millennia ago, with the ascent of Upper Paleolithic and Neolithic peoples, whose improved tools, dense populations, and deadly efficiency in the pursuit of wildlife inaugurated the current extinction spasm.[1]

The good news, Wilson said, is that scientists have found that "the biosphere is far richer in diversity than they had originally thought." The bad news is that this diversity, over three billion years in the making, "is being eroded at an accelerating rate by human activity." We face a "bottleneck of

Guess what?

.ɪ. The fossil record tells us that most species have gone extinct over the eons and that mankind had nothing to do with their extinction.

.ɪ. In the last two decades, the number of mammals actually counted has *grown* by 25 percent.

.ɪ. In recent decades, a handful of known species extinctions have been vastly exceeded by the number of species that have been observed and described for the first time.

overpopulation and rising per capita consumption." He speaks of "our present crisis" and of "mass extinctions." Human beings are to blame.

Wilson has undergone a remarkable transformation. In the mid to late 1970s, he was demonized by the Left. His book *Sociobiology: The New Synthesis* argued that human behavior was to some extent controlled by genes. His was a moderate position, on the whole, compared with the later tendency of biologists to posit genes for everything under the sun. He was nonetheless vilified by Harvard progressives, who found his views on genetic determinism to be sexist and racist. A left-wing activist dumped a jug of ice water over his head when he spoke at the Smithsonian Institution.

In time, the culture broadly shifted to a position close to Wilson's. Biologists today talk blithely of rape genes without so much as a peep from the Cambridge communards, who have gone into retirement. Meanwhile, Wilson's sociobiology has been preserved, expanded upon, and relabeled "evolutionary psychology"—a field of study in which hunches about human nature are dressed up in quasi-scientific guise.

At the same time, without repudiating his earlier views, Wilson moved into more fashionable political terrain. All the correct environmentalist positions, from the destruction of the Brazilian rain forest to endangered species to overpopulation and "wasteful consumption" became his own. He demonized "technomania," saw the human race as planetary despoiler, and praised demonstrators at meetings of the World Trade Organization. He wrote, He wrote, in an over-the-top "Letter to Thoreau": "The natural world in the year 2001 is everywhere disappearing before our eyes—cut to pieces, mowed down, plowed under, gobbled up, replaced by human artifacts."[2]

Whereas the evolutionists of old were in thrall to the Victorian idea of progress, Wilson, one of our leading evolutionists, had become remarkably pessimistic about mankind. He had fallen into line with the intellectual fashion of the day.

How much truth is there to these claims of vanishing species, loss of biodiversity, the natural world "disappearing before our eyes"? Wilson's evidence is murky, and finally unpersuasive. His claims about the much higher rates of extinction today—maybe 10,000 times higher—are based on concealed assumptions that do not hold up to scrutiny.

In the old biology textbooks, before environmentalism became a political cause, we were told that 99 percent of all species went extinct before man appeared on the scene. Extinction was to be expected. Darwin devoted a chapter to it in *On the Origin of Species*, and, far from regarding it as a tragedy, saw it as playing an essential role in evolution. Natural selection "almost inevitably causes much Extinction of the less improved forms of life," he wrote. (We're talking Survival of the Fittest, remember. So the less fit fall by the wayside.) As late as the 1960s, evolutionary biologists were inclined to believe that all species went extinct eventually.

How many species, how many extinguished?

The number of species actually counted "comes to somewhere between 1.5 million and 1.8 million right now," Wilson says, "but the estimates of the actual numbers out there range up to 100 million species and beyond." His numbers are vague—"a roster of ten million or more are still with us," he says.[3] So we don't know how many species exist to within maybe two orders of magnitude. Fewer than two million counted; maybe a hundred million in all. "We have only begun to explore life on Earth," he allows.

But this casts doubt on the claimed losses. Imagine a school principal who says that large numbers of students are playing hooky. Asked how many are on the school rolls, he says 1,600 have been counted, but the true number could be 16,000, or even 160,000. We would expect an accurate count to precede any warning about the students' absenteeism.

Patrick Moore: Founder of Greenpeace

Patrick Moore left Greenpeace in 1986—the year they established a pension plan, he said. He was interviewed by the British magazine *New Scientist* in 1999. Here is an excerpt:

Q: *How has the environmental movement got it so wrong?*

A: The environmental movement abandoned science and logic somewhere in the mid-1980s, just as mainstream society was adopting all the more reasonable items on the environmental agenda. This was because many environmentalists couldn't make the transition from confrontation to consensus, and could not get out of adversarial politics. This particularly applies to political activists who were using environmental rhetoric to cover up agendas that had more to do with class warfare and anti-corporatism than they did with the actual science of the environment. To stay in an adversarial role, those people had to adopt ever more extreme positions because all the reasonable ones were being accepted.

Q: *But hasn't environmentalism always been about opposing the establishment?*

A: Environmentalism was always anti-establishment, but in the early days of Greenpeace we did not characterize ourselves as left-wing. That happened after the fall of the Berlin Wall, when a whole bunch of left-wing activists, who no longer had any role in the peace, women's, or labour movements, joined us. I would go to the Greenpeace Toronto office and there would be an awful lot of young people wearing army fatigues and red berets in there.

The fossil record tells us that most species have gone extinct over the eons—but we know that mankind had nothing to do with that. The last great extinction, in which the dinosaurs died off, was sixty-five million years ago. Even in modern times, it is not possible definitely to attribute any given extinction to human activity. (A case could be made for the passenger pigeon—shot by the millions in the nineteenth century. The last specimen died in the Cincinnati Zoo in 1914.) The woolly mammoth, not seen on earth for about 4,000 years, was surely not done in by humans. The giant moa, a flightless bird living in New Zealand until about three hundred years ago, most probably was—by Maori tribesmen.

There have been five great extinctions, and the political enviros of our time claim that we are now living through a sixth—caused by human rapacity.

A special "biodiversity" edition of the politically correct *National Geographic*, published in 1999, claimed that the "sixth extinction" is here and now. "Half of all species could be annihilated in the next century," the magazine claimed, without warrant. (Wilson has said the same thing.) Biologist Stuart Pimm was quoted: "It's not just species on islands or in rain forests or just birds or big charismatic mammals. It's everything and it's everywhere. It is a worldwide epidemic of extinctions."[4]

Patrick Moore, a co-founder of Greenpeace and its president from 1977 to 1979, tried to find the evidence for these dramatic claims. A graph in *National Geographic*, showing the number of taxonomic families on earth over the past 600 million years, depicted a steady increase despite the previous waves of extinction. But when it reached the present day, it turned abruptly downward, indicating the losses due to our own "mass extinction."

Moore wrote to *National Geographic*, asking the magazine to identify any families known to have gone extinct in recent times. He himself did not know of "any families of 'beetles, amphibians, birds, and large mammals' that have become extinct as implied in the text." The reply came from a researcher who had worked on the article. She thanked him for "sharing" his thoughts "on this complicated and controversial issue" but offered no answers. She wrote:

> Rest assured that . . . the many members of our editorial team . . . worked closely with numerous experts in conservation biology, paleobiology, and related fields. The concept of a "sixth extinction" is widely discussed and, for the most part, strongly supported by our consultants and other experts in these areas, although specific details such as the time frame in which it will occur and the number of species that will be affected continues to be debated.[5]

Nowhere in the *National Geographic* article was there any recognition that the "sixth extinction" is controversial. It was simply presented as a known fact. It was also clear from the reply Moore received that the "mass extinction" was actually still in the future.

"In other words," Moore wrote, "there is no evidence that a mass extinction is actually occurring now, even though the article plainly implies that it is. The reply also refers to the sixth extinction as a 'concept,' implying that it is just an idea rather than a proven fact."

This is the level to which environmentalism has sunk. It is astonishing that it should have taken root within mass-circulation, once mainstream magazines such as *National Geographic*, and that it should have been embraced by scientists of E. O. Wilson's caliber.

Political science

Scenarios of catastrophe began spreading out from Washington a generation ago. In 1970, Dr. S. Dillon Ripley, secretary of the Smithsonian Institution, predicted that somewhere between 75 and 80 percent of all animal species would be extinct by 1995. A few years later, the famously unreliable Paul Ehrlich, whose prediction that sixty-five million Americans would die of starvation in the 1980s now seems comical, made a prediction comparable to Ripley's. The *Global 2000 Report*, published in 1980, predicted that 15 to 20 percent of all species would be extinct by the year 2000.

Examining these figures later, Aaron Wildavsky and Julian Simon concluded they were "pure guesswork." But that was putting it politely. They were mainly an expression of the misanthropy of the academy. Ehrlich's erroneous predictions have been matched by his receipt of as many prizes and awards. He was rewarded not because he was wrong, of course, but because he accurately perceived what so many intellectuals and academics want to hear: bad things about human beings and ominous things about the planet.

The principal source for the 40,000 species lost predictions turned out to be a 1979 book, *The Sinking Ark*, by British scientist Norman Myers—"a major scholar in biodiversity studies," according to Wilson. Myers somehow managed to persuade people that 40,000 species go extinct every year. This figure was later criticized by Bjørn Lomborg, another former Greenpeace member who started to think for himself. His book *The Skeptical Environmentalist*, first published in English in 2001, was a turning point.[6]

A comprehensive challenge to the Green faith, Lomborg's book was taken seriously by journalists who were getting a little bored with the doom-saying that never seemed to pan out, and it received wide publicity.

Perverse Consequences

PIG

"Under the Endangered Species Act, the owner must sacrifice any use of the property that federal agents believe might impair the habitat of the species—at the owner's expense. Furthermore, if the owner either harms the species or impairs its habitat, severe penalties are imposed. The perverse incentives created by the law may well lead an owner to surreptitiously destroy that animal or plant—or any habitat that might attract it."

Richard L. Stroup
"The Endangered Species Act: A Perverse Way to Protect Biodiversity," PERC Viewpoints, April 1992

It was also attacked by science journals. Lomborg was regarded more as a turncoat than a truth-teller. When he spoke at Oxford University, he had a pie thrown in his face, demonstrating, alas, the level at which some enviros are most comfortable debating the issues.

Here is how Myers had arrived at his 40,000 extinctions per year. He wrote:

> Let us suppose that, as a consequence of this man-handling of the natural environments, the final one-quarter of this [twentieth] century witnesses the elimination of one million species, a far from unlikely prospect. This would work out, during the course of twenty-five years, at an average rate of 40,000 species per year.[7]

And that was it. Matt Ridley, another reformed environmentalist turned critic, commented: "No data at all—just a circular assumption: If 40,000 species go extinct a year, then 40,000 species go extinct a year. *Q.E.D.*"[8]

In its own low point, *Scientific American* devoted eleven pages to attacking Lomborg, but only two trivial errors were found in his work.[9] Nonetheless, "the great and the good of greendom" competed to vilify the Dane, as Ridley put it. E. O. Wilson decried Lomborg's "scam," and was dismayed by "the extraordinary amount of scientific talent that had to be expanded to combat it in the media."

But it was precisely the non-deployment of scientific talent that had allowed the exaggerated claims of vanishing species to flourish in the first place.

Harper's magazine joined the chorus in 1998, with a long article titled "Planet of the Weeds." David Quammen, another award-winning scribe, relished the prediction of "conscientious biologists" that we are "headed into another mass extinction, a vale of biological impoverishment." The survival of the worst, is what it looked like. Wildlife will consist mostly of "pigeons, coyotes, rats, roaches, house sparrows, crows, and feral dogs." And lots of invasive species, too: zebra mussels, Asian gypsy moths, kudzu, and boll weevils aplenty! Worst of all, of course, there would be more humans beings than ever—the source of the trouble.[10]

Not to be outdone, the World Wildlife Fund announced at a 1996 media conference in Geneva that 50,000 species go extinct each year due to human activity. The main cause was said to be commercial logging. The story was carried around the world by the Associated Press. Once again, Patrick Moore tried to unearth the details. He pestered the World Wildlife Fund to name some of the species that had become extinct in this way.

"They have not offered up a single example as evidence," he wrote. "In fact, to the best of our scientific knowledge, no species has become extinct in North America due to forestry." He added:

> Where are these 50,000 species that are said to be going extinct each year? They are in a computer model in Edward O. Wilson's laboratory at Harvard University. They are electrons on a hard drive, they have no Latin names, and they are in no way related to any direct field observations in any forest.[11]

Wilson himself did some of the original research on which these extinction claims are based. In the 1960s, he and Robert MacArthur hired exterminators to destroy living things on small mangrove islands off the coast of Florida. They actually tear-gassed the critters! Sometimes they used chain saws to saw off portions of tiny islands. Then they estimated the percentage of species that disappears when habitat is lost.

Their answer: Destroy 90 percent of the habitat, and you lose 50 percent of the species. This rule of thumb—Wilson sometimes calls it an Iron Law—was then extended from islands the size of a house to the millions of square miles of the Brazilian rain forest. The press corps soon became a publicity machine for this campaign, and the new threat to "biodiversity" was on the front pages in no time.

It does not follow that a species that is "lost" on a particular tear-gassed island has gone extinct, of course. But confusion between "loss of diversity" and extinction soon set in. If a species is defined narrowly enough—and that is what happened—the extrapolation from "loss" to extinction is inconspicuous.

Wilson's experiments gave inconsistent results when repeated. Writing in the journal *Nature* in 1996, Lawrence Slobodkin concluded that other studies had shown that Wilson's species/area formula was "useless for explaining or predicting actual cases."[12]

Estimates of extinction rates also depend on the assumed size of a "minimum viable population." In the jargon, this is known as MVP. Ecologist Rowan Martin has pointed out that the trend has been to increase this number. We are "in an era of MVP hyperinflation," he says. MacArthur and Wilson had suggested that between twenty-five and fifty are needed to sustain

As Mark Twain Said: "The reports of my death are greatly exaggerated"

Scientific American recently wrote that the ivory-billed woodpecker is not the only species to be brought back from oblivion. Just a few days after the ivory-bill news, it reported:

"The Nature Conservancy announced the discovery in Alabama of three snails listed as extinct. A few weeks later, botanists at the University of California at Berkeley reported finding the Mount Diablo buckwheat, a tiny pink flowering plant that had not been seen since 1936. At least twenty-four species of other presumed or possibly extinct plants, insects, and other organisms have been found during natural heritage surveys in North America since 1974, according to Mark Schaefer, president of NatureServe, a nonprofit conservation group based in Arlington, Virginia."

"When Extinct Isn't," *Scientific American*, August 2005

a population. More recently, however, the minimal viable size is said to range from 10,000 to one million. "If these figures were applied to mountain lions, an area larger than the United States would be required for effective conservation," Martin wrote.[13] An expert on the fauna of southern Africa, Martin is unaware of any species in that vast area that has gone extinct in the last thirty-five years, and he suspects that some authors in this field have been "deliberately alarmist in their estimates."

Actual cases

What species are known to have gone extinct? The number actually observed during the past century in well-studied groups such as birds and flowering plants is small. Wilson says the annual rate of extinction may be no more than ten per million. If so, and with 10,000 bird species known to exist, it would imply that ten bird species have gone extinct in a century.

Charles C. Mann and Mark L. Plummer say that five North American birds disappeared as a result of hunting and forest clearing east of the Mississippi and around the Great Lakes. The first on that list was the ivory-billed woodpecker—since found to be alive and well in a swamp in eastern Arkansas. It had gone missing for sixty years, but recently was both videotaped and tape recorded. The other four birds listed as extinct are the passenger pigeon, the Carolina parakeet, the heath hen, and Bachman's warbler.[14]

In *Facts not Fear: A Parent's Guide to Teaching Children about the Environment*, Michael Sanera and Jane S. Shaw examined the well-studied record of wildlife in North America and found that since the early nineteenth century, "only a few animal species actually became extinct." One or two, notably the bison, or American buffalo, were severely reduced in numbers and genuinely endangered. It was saved in a private conservation campaign that began in the 1880s. The few remaining bison were trapped and moved to private land. Their numbers have now multiplied to over 350,000.[15]

Pessimistic forecasts, based on vague extrapolations, contrast sharply with contemporary observation and description of species.

- In the last two decades, the number of mammal species actually counted has grown by 25 percent, from 4,000 to 5,000. (Six new mammals were found in three weeks). No fewer than thirty-eight new species of monkey have been discovered since 1980.[16]
- In 2003 a new species of mangabey, a primate larger than a baboon, was discovered in Tanzania. Researchers were "stunned" because Tanzania is considered one of the biologically best-known countries in Africa. It showed just how little the experts know.
- A new species of macaque was discovered in India in 2004—the first new macaque discovered in a hundred years. "Few would have thought that with over a billion people and retreating wildlands, a new large mammal species would ever be found in India of all places," commented the Wildlife Conservation Society.
- From 1985 to 2001, the global number of amphibian species known grew by one-third. About 2,000 new species of flowering plants, and perhaps 15,000 species of all types, are added every year.

In fact, what may be going extinct is logic. When the new mangabey was discovered in Tanzania, a curious anomaly became apparent. The new species, a large mammal found in a well-explored area, had obviously been hard to find. Almost by definition, therefore, it was "endangered."[17] Consequently, a new species discovered could logically also be added to the endangered species list.

In summary, we have seen in recent decades a handful of known species extinctions, vastly exceeded by the number of species that have been observed and described for the first time.

Political taxonomy

In his description of bird species, Wilson notes at one point that an "unexpected revolution in field studies opened the census to a flood of new candidate species. Experts had come to recognize the possible existence of large numbers of sibling species."

This "unexpected revolution" was written into the Endangered Species Act of 1973, which defines species to include "any distinct population segment of any species of vertebrate fish or wildlife." The number of designated species promptly rose, but it became much easier for the doomsayers to claim that a given species had gone extinct, was on the verge of doing so, or would do so if the bulldozers in a particular place weren't stopped. Taxonomy itself was quietly politicized.

In *The Future of Life*, Wilson says that 16 of the 262 mammals native to Australia "are known to have vanished since the arrival of the European settlers."[18] That sounds bad until you read the list. Here are five: the Darling Downs hopping mouse, the big-eared hopping mouse, the short-tailed hopping mouse, the Alice Springs mouse, and the long-tailed hopping mouse. When species are so narrowly defined, some can be declared extinct with the stroke of a pen. Mice are still hopping in Australia, but maybe not on Darling Downs.

To be sure, there have always been disagreements between "lumpers" and "splitters." Two specimens are said to belong to different species if they cannot interbreed in "natural conditions." But no one is sure what "natural" means, nor can a failure to reproduce be confidently attributed to a biological inability to do so. Maybe the pair under observation weren't in the mood? So there is an arbitrary element to classification.

A Book You're Not Supposed to Read

The Skeptical Environmentalist by Bjørn Lomborg; Cambridge: Cambridge University Press, 2001.

But this has been exploited, politically, by scientists who know what they are looking for.

If a variety newly designated as a species can be shown to occupy a small enough range, any disturbance of that range can be said to threaten it. The Competitive Enterprise Institute's Fred Smith, who has spent years studying the stratagems of environmentalists, says that groups who use environmentalist arguments to cloak an anti-growth agenda operate with this maxim: "Let me define the habitat and I'll find you the endangered species."

Mussels are what the environmentalists "are playing games with now," Smith said. Because they change in outward appearance every three or four miles, it's easy to argue that different species are endangered all along the shoreline. But it has never been established that they are so different that they are incapable of reproducing or really deserve to be called different species.

Antitrust policy uses an analogous strategy, Fred Smith said. The ambitious prosecutor knows that if he is allowed to define the market, he can easily find monopolies and then swing into action, "doing his job" as prosecutor. Not long ago, a defendant was accused of "monopolizing the market for ready-made pizza dough inside Salt Lake City limits," Smith said. Control over the way the market was defined had conjured up a monopoly out of thin air and made life chancy for dough makers.

Issues of property

Something environmentalists don't want to hear is that property rights are intimately connected with the survival and flourishing of species. In the years ahead, with continued population growth (in Africa especially),

Greens will have to come to grips with an issue that many of them would rather not face: the configuration of property rights in whatever terrain they are interested in. Some would prefer to see the world population shrink (Ehrlich says a reduction of two-thirds would be "optimal" for the earth.) But that is wishful thinking, and it is not going to happen. Private property was about the last thing many Greens had on their minds when they went into the environmental business—and it *is* a business—but it turns out to be the unavoidable solution to their problems. Here are several points for them to consider:

- Valued animals in communal terrain where property rights are undefined are ipso facto endangered. A tribesman who takes one such animal, whether to eat, skin, or milk, experiences the benefit but not the cost of the communal arrangement. His incentive is to take one more, and then another, before someone else gets there first. The argument was spelled out by Garrett Hardin in a famous article called "The Tragedy of the Commons," published in *Science* in 1968. Environmentalists have shown little sign that they have even heard of the argument. Private owners, in contrast to communal dwellers or tribesmen, experience both the benefits and the cost of ownership, can exclude intruders, and so can exercise good stewardship.

- The American buffalo was almost exterminated because of the poor incentives generated by communal rights of those living in the central plains. The animal was saved because the remaining specimens were moved to private ranches. In contrast, the great mau, a flightless bird in New Zealand, was lost because land there in the seventeenth century was unowned. Maori tribesman hunted them all down, and that was the end of the mau.

- One famous species, the tiger, is threatened in the wild because the same problem exists in many parts of the tropical world today. There are thought to be no more than 5,000 tigers at large in the world. But the animal is not endangered, because an estimated 10,000 tigers are kept in private homes and cages in the U.S. alone. Fully grown, they eat twenty pounds of meat a day and make for dangerous pets. Tiger kittens can be sold, however, and so they do have commercial value. Owners have an incentive to rear them. (Good news? You would think so, but racially fussy enviropurists sniff at these mongrel big cats with their uncertain "blood lines.")

- Commercial value, therefore, helps to preserve large animals who would otherwise be hazardous and costly to keep. The elephant is the best example. Its commercial value lies in the ivory of its tusk. The international ban on the sale of ivory, accepted by the U.S. in 1989, was therefore hazardous to elephants, just as a ban on the sale of beef here would be hazardous to millions of cows. Deprived of the value of the ivory, African villagers are apt to regard elephants as a nuisance—"giant rats," as Fred Smith once put it. They can tear off the tops of village grain stores and consume an entire season's food supply on the spot.

- The Endangered Species Act is itself hazardous to endangered species, because if such a species—birds in particular—appear on private property, that property is defined as "critical habitat" and immediately comes under the jurisdiction of the Fish and Wildlife Service of the U.S. Department of the Interior.

The colloquial phrase for this is "shoot, shovel, and shut up."

California congressman Richard Pombo, the chairman of the House Resources Committee, put it more politely: "If a species is listed in a particular area, farmers and ranchers start managing the land so it doesn't attract the wildlife. They remove the possibility that they may have an endangered species on their property. That way they don't have a problem."

Pombo addresses the problem in a reform of the Endangered Species Act that passed the House of Representatives in September 2005. But "prospects for Senate passage are cloudy at best," says the *New York Times*.[19] On this issue, the Greens still seem to control public opinion. The problem is that they are incapable of recognizing that their good intentions may have unanticipated or untoward results. Human nature, which they disapprove of so much, rarely enters into their calculations.

Chapter 7

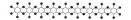

AFRICAN AIDS
A POLITICAL EPIDEMIC

AIDS in Africa beats even global warming as a zone where politics trumps science. Independent research is not welcome—and is even punished. Almost overnight, in 1985, millions of Africans were said to be dying of AIDS, and on cue journalists fell into line. There were a few exceptions. Celia Farber wrote two good articles for the music magazine *SPIN*, questioning the whole story. South Africa's Rian Malan wrote a great article for *Rolling Stone*. Liam Scheff, a young writer whose work has been confined to the alternative press, did some good investigative work. But overwhelmingly, media coverage of AIDS in Africa was slavish.

Government press releases were published without question. This was an emergency, and doubt itself was viewed as a luxury. The lives of millions were at stake. In the U.S. press, a general credulity about African AIDS has endured for twenty years—a long time in journalism.

The full story has never been told. In fact, what the public has *not* been told is the most important part of the story.

First, a few numbers. The international agency UNAIDS claims that, at the end of 2004, about twenty-five million people in sub-Saharan Africa were living with HIV, the virus that is said to cause AIDS. (This is down from an earlier estimate of thirty million.) In this vast region, there

Guess what?

.∴. The AIDS pandemic in Africa was invented at a meeting in the Central African Republic in 1985.

.∴. Many HIV tests were conducted at South African prenatal clinics. Pregnancy is just one of many conditions that produce a "false positive" on the test.

.∴. African AIDS countries were flooded with condoms, U.S. aid, and UN attention. But confounding the predictions, the population of sub-Saharan Africa *increased* by 299 million.

How to Save a Marriage

"Several years ago I acquired what was generally regarded as a leprous obsession with the dumbfounding AIDS numbers in my daily papers. They told me that AIDS had claimed 250,000 South African lives in 1999, and I kept saying this can't possibly be true. What followed was very ugly—ruined dinner parties, broken friendships, ridicule from those who knew better, bitter fights with my wife. After a year or so, she put her foot down. Choose, she said. AIDS or me. So I dropped the subject, put my papers in the garage, and kept my mouth shut."

Rian Malan, "AIDS in Africa, in Search of the Truth," *Rolling Stone*, November 22, 2001

were (supposedly) three million new HIV infections in 2004, and over two million AIDS deaths.

By contrast, the United States has seen a cumulative total of about 525,000 AIDS deaths since 1981, when the disease was first diagnosed. In South Africa, in just one year (2003), 370,000 people supposedly died of AIDS. There were 18,000 American AIDS deaths in the same year. In other words, one-twentieth the deaths occurred in a country with over six times the population.

We must first look at the definition of AIDS. It is a complex subject but crucial. There have been at least four definitions over the years, and if an official from the U.S. Centers for Disease Control and Prevention (CDC) in Atlanta, Georgia, promises to fax all the definitions to you, be sure to have lots of paper in your fax machine!

Boiled down, it goes like this. The human immunodeficiency virus (HIV) eats away at the T-cells of the body's immune system, thereby exposing it to infections. Twenty-six diseases are now on the list of these "opportunistic" infections. Some of them are not actually infectious—Kaposi's sarcoma and cervical cancer, for example. Others are—tuberculosis, herpes, pneumonia, and candidiasis among them. So, if you have one of these diseases, *and* you are HIV positive, *and*, in time, your T-cell count dips below a certain level, then you have AIDS.

In Africa, however, the WHO—under the supervision of the CDC—put together a quite different "clinical case definition" of AIDS. It really is very different, in three crucial respects.

- None of the opportunistic diseases has to be present.
- No HIV test has to be conducted.
- No T-cells are counted.

AIDS redefined

This redefinition was worked out in October 1985, at a meeting in Bangui, the capital of the Central African Republic. About sixty officials were present, including Americans from the CDC who had organized the meeting. A document was produced, and the curious may view it on the web. Titled "Workshop on AIDS in Central Africa," it spells out the medical conditions considered sufficient to identify an AIDS case in Africa. (To view it, type "Bangui1985report" without spaces into your search engine.)[1]

What prompted the meeting was the claim that AIDS or something resembling it had "recently appeared" in hospitals in Zaire and elsewhere. The "biological features of AIDS in Africa" had to be identified. That would not be easy, however. "Adequate laboratory facilities are often lacking," as those in attendance were reminded. Instead, a "surveillance" definition was needed. It had to be "simple, universally applicable and usable by all health service personnel."

Given these limitations, here is what the participants finally agreed was needed to count AIDS cases in Africa. Four "major" symptoms had been found to be associated with AIDS in the Western world:

- Weight loss of 10 percent or more
- Pronounced weakness or lack of energy (called "asthenia")

- Diarrhea lasting for more than a month
- Fever, either prolonged or intermittent

In addition, several "minor" symptoms were often found, among them:

- A cough persisting for more than a month
- Chronic ulcerative herpes infection
- Swollen glands (called "generalized adenopathy")

From these symptoms, a new "definition of AIDS in adults in Africa" was derived by mixing and matching from the two lists:

"AT LEAST THREE OF THE FOUR MAJOR SYMPTOMS ASSOCIATED WITH ONE OF THE MINOR SYMPTOMS" [caps in original].

That was it. There was no mention of HIV.

Overnight, millions of Africans now had AIDS, by these criteria. The definition was so broad that "almost anyone in any African hospital could be said to have it," says Rian Malan. Let's say you're in the Congo. You go to a doctor because you're feeling weak. You've lost weight and have had a recurring fever for a few weeks and a persistent cough. Doctors are now free to say that you have AIDS. For a child, all they need is weight loss, diarrhea, and a cough.

In the document's remarkable final page, "Clinical Diagnosis of AIDS," points were awarded for the various conditions. A note at the bottom explained that "the diagnosis of AIDS is established when the score is 12 or more."

Here are some of the points awarded: weight loss—4; protracted feeling of weakness—4; repeated fever—3; chronic diarrhea—3; cough—2; swollen glands—2; "neurological signs"—2.

In Africa, "as in the United States," the report stressed, "sexual transmission is the main way in which AIDS is spread." No evidence was offered for that sweeping claim. Nonetheless, "in the absence of treatment

Clinical Diagnosis of AIDS in Africa

Exclusion criteria

1. Pronounced malnutrition
2. Cancer
3. Immunosuppressive treatment

Inclusion criteria with the corresponding scores

Important signs	Score
Weight loss exceeding 10% of body weight	4
Protracted asthenia	4

Very frequent signs	
Continuous or repeated attacks of fever for more than a month	3
Diarrhoea lasting for more than a month	3

Other signs	
Cough	2
Pneumopathy	2
Oropharyngeal candidiasis	4
Chronic or relapsing cutaneous herpes	4
Generalized pruritic dermatosis	4
Herpes zoster (relapsing)	4
Generalized adenopathy	2
Neurological signs	2
Generalized Kaposi's sarcoma	12

THE DIAGNOSIS OF AIDS IS ESTABLISHED WHEN THE SCORE IS 12 OR MORE

SOURCE: "Workshop on Aids in Central Africa," Bangui, Central African Republic, 22 to 25 October 1985, World Health Organization, 15.

or of a vaccine, health education aimed at changing sexual behavior is an essential means of controlling AIDS."

Finally, the media was enlisted to get the word out on the threat of AIDS in Africa:

"The mass media should be urged to play a part in this health education. Media personnel could receive training for this role."

The media duly reported whatever the authorities told them to report. Evidence gleaned from sick patients in African hospitals, physically resembling American AIDS patients, became the harbingers of a world pandemic. Pictures of emaciated Africans lying on cots were transmitted, printed, and reprinted. The sub-Saharan continent was flooded with condoms, pamphlets, and AIDS educators. Money flowed in by the boatload from aid agencies all over the world.

The new AIDS definition was reported in WHO's *Weekly Epidemiological Record*,[2] in CDC's *Morbidity and Mortality Weekly Report*, and in *Science*.[3] But it has not been visibly reported by the major news media in the United States.

In 1999, *Village Voice* reporter Mark Schoofs spent six months in Africa and wrote an eight-part series on AIDS.[4] While there he contracted malaria, and it is possible that by the African definition he had "AIDS" himself at that point. But he didn't mention the redefinition and possibly didn't know about it. He won a Pulitzer Prize and was hired by the *Wall Street Journal*.

The case of the *New York Times* is instructive. The Bangui definition was not reported by the paper, and that meant television could be counted on not to report it, either. This is why the vast majority of the American public has never heard of it.

From the beginning, in 1981, the *Times*'s major AIDS reporter has been Dr. Lawrence K. Altman. He wrote what may have been the first article on AIDS for a national newspaper. It appeared under the now politically incorrect headline, "Rare Cancer Seen in 41 Homosexuals."[5] But that was all soon to change. Heterosexual transmission was established by the

Bangui definition. The conditions that define AIDS in Africa occur in women as often as in men. Therefore AIDS was no longer primarily a homosexual disease.

Altman is a former employee of the CDC and sits on an advisory board that administers a CDC fellowship program. He graduated from the agency's Epidemic Intelligence Service in 1963 and became editor of the CDC's *Morbidity and Mortality Weekly Report*. Later he was chief of the U.S. Public Health Service's Division of Epidemiology in Washington.

In November 1985, Altman wrote two huge stories based on reporting from central Africa. One story included a box headlined "African Meeting Addresses Problems of AIDS." It was the Bangui meeting, but he said no more about the new definition than this:

> They will work to create a hospital surveillance system to determine the extent of AIDS in central Africa. Most African countries lack effective health-care reporting systems and the attempt to set up one for AIDS would be designed to detect trends rather than precise case counts.

No further details were forthcoming, particularly that HIV was no longer necessary for an AIDS diagnosis. What followed in the thousands of news media headlines in the years to come was a disservice to public health, because these alleged trends were construed as precise case counts. The numbers soon became wildly inflated and subverted epidemiology. Now it could be inferred that men and women were equally at risk. This was the message that Altman conveyed in his front-page lead:

> Kigali, Rwanda: AIDS appears to be spreading by conventional sexual intercourse among heterosexuals in Africa and is striking women nearly as often as men, according to researchers here.[6]

The Bangui meeting was organized by a CDC official named Joseph B. McCormick. His goal, he said, was to establish a diagnostic definition of

AIDS for use in countries not equipped to do blood tests. The CDC also persuaded the WHO in Geneva to set up its own AIDS program. It was a triumph of redefinition. Sick Africans in hospitals who looked superficially like American AIDS patients had persuaded American officials that AIDS now existed in Africa.

McCormick later co-authored a book, *Level 4: Virus Hunters of the CDC*. First published in 1996, it was reissued with new material in 1999. As to the Bangui definition, McCormick wrote:

> We needed a clinical case definition—that is to say, a set of guidelines a clinician could follow in order to decide whether a certain person had AIDS or not. This was my major goal. If I could get everyone at the WHO meeting at Bangui to agree on a single, simple definition of what an AIDS case was in Africa, then, imperfect as the definition might be, we could actually start to count the cases, and we would all be counting roughly the same thing.[7]

That goal was achieved. "The definition was reached by consensus, based mostly on the delegates' experience in treating AIDS patients." The authors added that it has been "a useful tool in determining the extent of the AIDS pandemic in Africa, especially in areas where no testing is available."[8]

There was also a political motive. McCormick had earlier telephoned a disbelieving assistant secretary of health, Dr. Edward Brandt, a reactionary "Reagan appointee," according to McCormick, who was wedded to the "gay plague" interpretation of AIDS. So McCormick needed to enlighten him, as a matter of some urgency. ("This was the Reagan era," he reminds readers.) What the CDC people, with their newfound African strategy, planned to tell Washington was that "AIDS was a plague all right, but that no one was immune."[9]

The extension of the AIDS epidemic into Africa made life easier for the CDC. For one thing, infectious epidemics normally break out evenly between the sexes; viruses are not supposed to discriminate by sex. (In the U.S. today, however, over half of the new HIV infections are diagnosed among black men, so this virus apparently discriminates both by sex and by race.)

In the 1980s, although the disease had remained stubbornly confined to the (overlapping) risk groups of promiscuous homosexuals and drug addicts, heterosexual transmission of the virus was confidently predicted. For example, Oprah Winfrey told her audience that one in five heterosexuals would be dead of AIDS by the 1990s. Surgeon General C. Everett Koop referred to the "heterosexual AIDS explosion." But by mid-1986, the CDC was still reporting that 93 percent of confirmed AIDS cases were male.

After Bangui, promiscuous African "truck drivers," traveling from city to city, became the unwitting agents of heterosexual transmission. Hollywood lifestyles were casually imputed to Africans. This was an epidemiological necessity, because careful U.S. studies had already shown that at least a thousand sexual contacts are needed to achieve heterosexual transmission of the virus.

Soon enough, with the huge but unverified numbers from Africa swamping the verified numbers from the U.S., the worldwide AIDS statistics showed an equal division between the sexes. So two obstacles, epidemiological and political, were removed at one stroke at Bangui. The new AIDS mantra became: "We are all at risk." AIDS funding was destined to soar.

False positive

The Bangui definition made it easy for health bureaucrats to diagnose AIDS easily, evidently with the hope that millions of cases would soon be reported to the WHO. But this did not happen. The WHO's *Weekly*

Epidemiological Record reported at the end of 1999, for example, that for the whole of Africa a cumulative total of 794,000 cases of AIDS cases had been reported. In November 2000, the same WHO publication reported a cumulative total of only 12,825 AIDS cases in South Africa since 1982.[10]

These numbers were still not earth-shaking, especially when broken down by country. Egypt, for example, with a population over 65 million, reported a cumulative total of 215 AIDS cases over 17 years. Soon, however, the figures were enormously augmented. HIV tests were conducted at prenatal clinics in various African countries, but primarily in South Africa. The numbers found to be "HIV positive" in these tests were then extrapolated to the whole country and then added to the "clinic surveillance" cases. By this stratagem, it was found that the number of South Africans "living with HIV" in 2000 (to choose just one year) had grown to 4.2 million.

The underlying problem here, rarely reported, is that HIV tests react to lots of conditions in addition to HIV. The antibody tests react to certain proteins that are not specific to HIV antibodies. Parasites that cause malaria, in particular, confound the test. Rian Malan reported a test published in a scientific journal in which a special preparation that absorbed the malaria antibodies was added. At that point, 80 percent of the suspected HIV infections vanished, he reported.[11]

UNAIDS head Peter Piot himself accepts that the immune systems of Africans are often "in a chronically activated state, associated with chronic viral and parasitic exposure." In South Africa's pregnancy clinic surveys, a single ELISA (Enzyme Linked Immunoabsorbent Assay) test is deemed sufficient confirmation. Yet Abbott Laboratories, the manufacturer of the test, has warned that pregnancy itself is among the conditions that can yield a misleading result.

The Western Blot test, often used to "confirm" HIV infection, also reacts to other conditions. Flu, herpes, immunizations, hepatitis, blood transfusions, parasites, tuberculosis, and malaria are among the dozens of conditions that can give rise to false positives. The packet insert in an HIV/ELISA test from Abbott Laboratories, the world leader in HIV testing, contained this disclaimer in 1997: "At present there is no recognized standard for establishing the presence or absence of antibodies to HIV-1 in human blood."

Yet the main surveillance study for HIV prevalence in South Africa depends on a single ELISA test administered to pregnant Africans, never acknowledging that pregnancy is "one of seventy conditions known to trigger a 'false positive' result."[12]

Conditions that define AIDS in sub-Saharan Africa are caused by many germs, not just HIV. And the other diseases caused by these germs also produce a "positive" result in the HIV test. Charles Gilks wrote in the *British Medical Journal* that persistent diarrhea with weight loss can be associated with "ordinary enteric parasites and bacteria," as well as with opportunistic infection. "In countries where the incidence of tuberculosis is high," he added (and it is high in Africa) "substantial numbers of people reported as having AIDS may in fact not have AIDS." He concluded that the Bangui definition "is inherently unworkable and incorrect."[13]

Life and death

In 1998 the *New York Times* published a lengthy series of articles about African AIDS, titled "Dead Zones."[14] The articles were mostly veiled sermons against "social attitudes and gender relations," "stigma," "silence," "superstition," and "conservative religious beliefs." Oddly, the series failed to report statistics from the one sub-Saharan country with respectable data on births and deaths—South Africa. All the published figures came from Geneva—the same old unreliable WHO estimates of HIV/AIDS. This made no sense, because the great majority of South African deaths are registered; elsewhere in sub-Saharan Africa, most are not.

Rian Malan did some investigating on his own in South Africa. According to the WHO estimates, South African AIDS deaths had more than tripled from 1996 to 1999. But the overall deaths reported by Pretoria had increased only by about 13 percent in the same period. It seemed likely, therefore, that a large number of deaths had simply been shifted over into the AIDS column.

Malan interviewed coffin makers in Cape Town. If the crisis reports were correct, they were surely doing a booming business. "All across Africa," the *New York Times* had reported (but without giving details), "the coffin business boomed."[15] But when Malan investigated, he found that entrepreneurs who were trying to sell cheap cardboard caskets had gone out of business. "People weren't interested," he was told. "They wanted coffins made of real wood."

> So I called the real wood firms, three industrialists who manufactured coffins on an assembly line for the national market. "It's quiet," said Kurt Lammerding of GNG Pine Products. His competitors concurred—business was dead, so to speak.
>
> "It's a fact," said Mr. A. B. Schwegman of B&A Coffins. "If you go on what you read in the papers, we should be overwhelmed, but there's nothing. So what's going on? You tell me."

Malan didn't know, so he investigated Johannesburg's derelict downtown, where coffin makers can be found. It was the same story. One likely place turned out to be "locked up and deserted. Inside I saw unsold coffins stacked ceiling high, and a forlorn CLOSED sign hung on a wire."

Hysteria about African AIDS reached a peak in 2000, when Vice President Al Gore took the issue to the UN Security Council. World Bank president James Wolfensohn said the epidemic was "more effective than war in destabilizing countries." It was even foreseen that Botswana's population might disappear. The toll was such that Africa was found to have ten million orphans. (We weren't told that in the new lexicon, an "orphan" in Africa is defined as someone under fifteen with *one* parent "missing," not necessarily dead.) The continent was heading for the abyss.

An earlier *New York Times* series on African AIDS, published in 1990, emphasized the need for condom distribution—as though that had been overlooked. The paper reported that USAID "has given seven billion condoms to developing countries." Since then, of course, billions more have been shipped.[16]

Africans could be forgiven for thinking that condoms are America's principal export. They may even be under the impression that our educated classes think that there are too many sub-Saharan Africans. Let's hope they don't see the recent newspaper column by CBS commentator Andy Rooney, who blurted out what may indeed be on the minds of some of our more hard-hearted compatriots.

Rooney said he would like to see more American aid spent on "reducing the number of Africans we're trying to feed. Their biggest problem is not a shortage of food, but a proliferation of people.... The birthrate in Africa is a disgrace and birth control information and condoms should be handed out before the food."[17]

Someone should tell Mr. Rooney about the U.S.-funded AIDS program. It is, above all, a condom distribution program. Perhaps Rooney did hear,

Health Conditions in Africa

"Look at AIDS from an African point of view. Imagine yourself in a mud hut, or maybe a tin shack on the outskirt of some sprawling city. There's sewage in the streets, and refuse removal is nonexistent. Flies and mosquitoes abound, and your drinking water is probably contaminated with feces. You and your children are sickly, undernourished, and stalked by diseases for which you're unlikely to receive proper treatment. Minor scourges such as diarrhea and pneumonia respond sluggishly to antibiotics. Malaria now shrugs off treatment with chloroquine, which is often the only drug for it available to poor Africans."

Rian Malan, "AIDS in Africa: In Search of the Truth," *Rolling Stone,* November 22, 2001

however, that all the dire predictions about the AIDS-caused population collapse in Africa turned out to be false.

Quiet population explosion

In the end, Rian Malan defied his wife, retrieved his papers from his garage in Cape Town, and went back to speaking out about the non-epidemic in Africa. Recently, he drew attention to a July 2000 prediction by the U.S. Census Bureau and USAID: "By the year 2003, Botswana, South Africa, and Zimbabwe will be experiencing negative population growth."

Malan commented:

> When 2003 rolled around, both Botswana and South Africa had conducted censuses, and the results were mortifying: both countries' populations were growing fairly rapidly. In South Africa, growth came very close to the level projected with no AIDS at all—even among young adult females, who were supposedly dying like flies of HIV infection. As usual, this confounding development was ignored by the craven lickspittles of AIDS journalism. The "experts" were allowed to reschedule the apocalypse for 2010, and the great ship AIDS sailed serenely onwards.[18]

Journalists, notably Laurie Garrett, who has specialized in predicting plagues that never arrive, have compared the African AIDS epidemic to the Black Death, in which perhaps one-third of Europe's population died in a few years. What are the numbers from Africa?

The UN Population Division estimates that the sub-Saharan population of Africa was 434 million in 1985—the year when the "AIDS pandemic" began. Recently, the Population Reference Bureau in Washington, D.C., published its "World Population Highlights." The sub-Saharan population in 2004 was 733 million, they said. Let's say that population remained the same in 2005. Since 1985, then, the population of sub-Saharan Africa has increased by 299 million people—a 70 percent increase.

Today, "sub-Saharan Africa and western Asia are the fastest-growing regions of the world," says the Population Reference Bureau.

The population of the United States is estimated to be 296 million. So in sub-Saharan Africa, in twenty years of plague, the population increased by more than did the entire population of the United States.[19]

Books You're Not Supposed to Read

AIDS: The Failure of Contemporary Science by Neville Hodgkinson; London: The Fourth Estate, 1996.

Inventing the AIDS Virus by Peter H. Duesberg; Washington, D.C.: Regnery Publishing, 1996.

Oncogenes, Aneuploidy and AIDS: A Scientific Life and Times of Peter H. Duesberg by Harvey Bialy; Berkeley: North Atlantic Books, 2004.

Post-colonial health

But there's something else that no one has wanted to talk about, not Rian Malan and certainly not the politically progressive rock stars Bono and Bob Geldof, who weighed in with Live8 concerts and were welcomed into the Oval Office. Always they had the same message: more, more, more money is needed.

In expanding the infectious epidemic to millions in Africa, as we have seen, the CDC solved first an epidemiological problem (infectious diseases don't discriminate by sex); and then a political obstacle (an equal-opportunity killer, putting us "all at risk," ensured that budgets would soar). But there was a third issue. This one was and still is bottled up by taboos.

In tropical Africa, a deterioration of the physical infrastructure swiftly followed the end of colonial rule. Sewage and sanitation crumbled. The issue was too awkward to mention because it would strongly suggest that Africans were better off—or at least in better health—under colonial rule.

Joseph McCormick, the CDC official who played such an important role in Bangui, had spent time in Kinshasa, Zaire, two decades earlier. He had lived there, or nearby, in 1965. It was then still "a city that worked—a marked contrast to what it's like these days," he wrote in *Level 4*. "Traffic coasted down wide, well-maintained boulevards flanked by palm trees. At night the city was brilliant with light, and when you turned on the taps, water would flow out."

Taps with running water, no less! Had the Belgians done something right after all? When McCormick returned in the 1980s, he was "struck by how much the place had deteriorated."

The president of the UN General Assembly, Jan Eliasson, said an interview with the *New York Times*: "You can break down figures that show that 300 million people south of the Sahara don't have clean water." He described the mother with infant who lights up with a smile when offered a bottle of water, knowing that "the only alternative is to walk for two or three miles and then only to get polluted water."[20]

This from London's *Daily Telegraph*, in July, 2005: On the outskirts of African cities,

> in vast shanty towns, human beings endure privation and squalor of a kind that disappeared from the rich world generations ago. Families live in shacks fashioned from cardboard boxes and twisted metal. Fetid heaps of rubbish are everywhere, sometimes dotted with scavenging vultures. People must defecate into plastic bags, tossing the results onto piles of refuse. Naked infants have nowhere to play, save for ditches running with sewage.[21]

And from the *Daily Mail*, also in July:

> African trypanosamiasis, also known as sleeping sickness, [is] one of several horrific diseases to have made a big resurgence

in south Sudan and other forgotten corners of Africa in recent years. Sleeping sickness, virtually eliminated in the 1960s, now afflicts up to 300,000 new victims in Africa each year and the figures are rising. Without treatment, the disease is fatal.[22]

It has been comforting to Western intellectuals to attribute the bad health of Africans not to the hazards and difficulties of self-government but to bad luck. The culprit was the human immunodeficiency virus—the "savage virus," as journalist Michael Specter has called it. And the well-known treatment for that had already been tested and approved by the AIDS activists of urban America: nucleoside analogs, protease inhibitors, highly active anti-retroviral treatment. Let them eat retrovir, nevirapine, AZT.

Lately, to be sure, there has been some improvement in the public understanding of the African plight. Perhaps news of the continued population increase is beginning to spread. To give them their due, the rock stars, Bono & Co., may have played a beneficial role. They have not been afraid to draw attention to the local corruption that has been far more widespread than the virus; they do mention the rulers who deposit aid money in their Swiss bank accounts. The aid agencies, in contrast, seek to increase their budgets above all else. Subordinate to the State Department, they place diplomacy before truth. The rock stars have different incentives, vanity admittedly among them. But they would also like to see improvement on the ground.

The recent attention paid to clean water is a welcome change of focus. Unlike AZT—originally designed as cancer chemotherapy and not good for the health of those who consume it—clean water and rebuilt sanitation systems will work wonders for the health of Africans.

Chapter 8

THE FOLLY OF DOLLY
CLONING AND ITS DISCONTENTS

T he news that scientists had cloned a lamb called Dolly caused a great sensation. It was "the most important story of the last two or three decades," according to a national magazine editor. Probably it was not so much the science as the misconceptions about cloning that had stimulated the interest. For some, the word has conjured up images of resurrection and even immortality.

Dolly was born in 1997. The cloning craze has since faded, and the supposed benefits have not materialized. Now bioengineers have latched onto a new (and related) enthusiasm: stem cells.

The cloning furor was in part inspired by science fiction, which unloosed many fantasies. In Ira Levin's *The Boys from Brazil*, Nazi doctor Josef Mengele tries to create clones of Hitler. It later became a movie starring Gregory Peck and Lawrence Olivier. The interstellar conflict in *Star Wars* was said to have started out as "clone wars." Naomi Mitchison, the sister of British biologist J. B. S. Haldane, wrote a book in which the survivors of a nuclear war control their own evolution by cloning only the best and brightest survivors. Cloning became eugenics, updated. A leading prophet of progressivism in science, Haldane invented the term "clone."

Some people believe that if we can clone ourselves, we may live on forever. *Yes to Human Cloning: Immortality Thanks to Science!* published

Guess what?

.•. Cloning creates serious abnormalities in almost all embryos, and the vast majority of the attempts to clone end in failure.

.•. Cloning has increased the doubts that genes alone contain all the "instructions" needed to make an animal.

.•. Cloning farm animals turned out to be so inefficient that its commercial promises will not be met.

123

in 2001, was written by a French journalist, Claude Vorilhon, who renamed himself Rael and formed the Raelian sect. He claims to be a direct descendant of extraterrestrials who created human life on earth through genetic engineering. In early 2004, his followers said that the first human clone—"Eve"—had been born in an "undisclosed country." That was an all too familiar story. The field of cloning has been replete with hoaxes.

Scientists were not responsible for these absurd claims; they were excited for other reasons. Before Dolly, it had been thought impossible to clone an animal from an adult cell—as Dolly had been. The nucleus of an udder cell was transferred to an egg from which the nucleus had been removed. The embryo was then grown in a surrogate sheep. As early as the 1950s, biologists had managed to insert the nuclei of adult frog cells into frog eggs and create tadpoles, but they had died at that stage. Dolly was the first animal successfully cloned from an adult cell.

All along, there have been bogus reports of cloning. In 1978, J. P. Lippincott published *In His Image: The Cloning of a Man*, and the author, David Rorvik, claimed it was factual. It became a bestseller, but both author and publisher were sued by an Oxford biologist whose work had been cited. In court, the book was declared "a hoax and a fraud."[1]

In 1979 a German researcher, Karl Illmensee, claimed that he and a colleague had cloned mice at the Jackson Laboratory in Maine. The *New York Times* reported this "first cloning of mammals" on page one (January 4, 1981). But the work proved to be unrepeatable. Illmensee left the country and took a position at the University of Geneva. Later, in 1983, faculty members signed a statement saying that he had "falsified ('faked') protocols including experiments that in reality had not been carried out."

Describing this episode in *Clone: The Road to Dolly and the Path Ahead*, Gina Kolata wrote that news of the incipient scandal spread from lab to lab in Europe and the United States. "But although newspapers like the *New York Times* and weekly magazines like *Time* and *Newsweek* had heralded Illmensee's claims to have cloned mice, they were strangely silent about the new charges that Illmensee's work might not be credible. A few short articles appeared but they hardly conveyed the dimensions of the controversy that was rocking the scientific community."[2]

But Illmensee admitted nothing in public and the fraud charges were unproven. Another researcher, Davor Solter, tried to replicate Illmensee's work but could not. His papers to that effect were published in *Cell* and in *Science* in 1984. "The cloning of mammals, by simple nuclear transfer, is biologically impossible," Solter concluded, defiantly. But that was the method later used to clone Dolly.

After Ian Wilmut of the Roslin Institute in Scotland reported the cloning of Dolly, he was supported by just about every scientist who had attempted cloning. Doubts nonetheless persisted. Norton Zinder, a professor of molecular genetics at Rockefeller University, said the paper by Wilmut and Keith Campbell, published in *Nature*, was "a bad paper." He was not charging fraud, but "lousy science, incomplete science."[3]

The udder cells used had been frozen in a vial for three years before the researchers used them for cloning, and the parental sheep was dead by the time Dolly was born. That made direct comparison impossible, although the DNA did match.

Another critic, Walter Gilbert, a Nobel laureate at Harvard, said there was no way of knowing for sure that Dolly had been cloned from an adult rather than an embryonic cell. Had there been a mixup? More of the udder cells were still in the freezer, so why not repeat the experiment? Wilmut said the process was too inefficient. He estimated it would take him over one thousand attempts to have a good chance of ending up with another live lamb. "That would cost half a million dollars and would be half of what I could do in a year."

Dolly had been the only pregnancy out of three hundred attempts to clone from udder cells. But he would try again, he said, using cells from a living animal.[4]

Dolly died at the age of six in 2003. Surprisingly, scientists don't know what the average life span of a sheep is. Some say seven years, some say fifteen. (Most are slaughtered and eaten before their time.) But few scientists today question whether Dolly was cloned from an adult. The process has been repeated with cats and cattle. And no one doubts that the cloning of animals turned out to be very much harder than anticipated.

When researchers finally cloned a dog from the adult cell of an Afghan hound, the South Korean researchers who reported the success "worked for nearly three years, seven days a week, 365 days a year, and used 1,095 eggs from 122 dogs before finally succeeding," Gina Kolata reported in the *Times*.[5] Rick Weiss reported in the *Washington Post* that multiple surgeries on more than one hundred anesthetized dogs, "and the painstaking creation of more than 1,000 laboratory-grown embryos led to the birth of just two cloned puppies—one of which died after three weeks."[6]

"Dollywood"

Obi-Wan: I have to admit that without the clones it would have not been a victory.

Yoda: Victory. Victory you say? Master Obi-Wan, not victory. The shroud of the dark side has fallen, begun the clone war has.

Star Wars: Episode II—Attack of the Clones (2002)

Doug Kinney #3: You know how when you make a copy of a copy, it's not as sharp as... well...the original.

Multiplicity (1996)

There have been other problems. According to Ian Wilmut, cloning seems to create serious abnormalities in almost all embryos. The genetic material in Dolly showed signs of premature aging, appearing to be older than Dolly herself.[7] It is doubtful that clones really are exact genetic copies. Some cloned mice, apparently normal when young, grow grotesquely fat as young adults, even when given exactly the same food as otherwise identical but uncloned mice.[8]

When a cat, called cc (for carbon copy) was cloned in 2002, its coat was not the same color as the two-year-old cat who provided the cell nucleus. Dr. Duane Kraemer, a member of the cat-cloning team at Texas A&M, said he was glad the clone didn't look like the original. "We've been trying to tell people that cloning is reproduction, not resurrection," he said.[9]

People do need to be told that, especially those who think their beloved pets can be brought back from the dead. It would be better and more humane to go to the pound.

The cloning of pets will cost $1 million, which ensures that it won't become a commercial enterprise. In fact, the rare successes of cloning have undermined the very dogma upon which it is based: that the DNA—which is what is transferred from the nucleus of the donor cell to the recipient egg—contains all the "instructions" that are needed to generate the cloned animal.

Jorge A. Piedrahita, a genomics professor at North Carolina State University's College of Veterinary Medicine, said that studies of pigs had convinced him that pigs that were clones of each other were no more alike than were conventional pigs with regard to food preferences, sleep habits, or levels of aggressiveness. "What was most fascinating is how important the environment is—that it really overrides the genetic similarities," he told Rick Weiss of the *Washington Post*.[10]

All of this has called into doubt the initial rationale for cloning: the best farm animals would be selected by profit-maximizing farmers and inexpensively replicated by cloning. It didn't work out. Cloned animals have turned out to be both expensive and non-replicas. Far simpler to breed them the old-fashioned way. The sheer inefficiency of cloning ensures that it cannot be sustained as a commercial enterprise.

A scientific experiment is one that can be repeated, and by that criterion animal cloning is more trial-and-error than science. Illmensee's cloning experiments could not be repeated, but neither, reliably, could others done more recently. Tanja Dominko tried to clone monkeys for years, but had no success after three hundred attempts. She succeeded in creating what she called a gallery of horrors—sometimes with cells that looked more like cancer cells than normal ones. A placenta with no fetus was the best she could do. She told Gina Kolata:

> If you want to make it [cloning] into something that will have commercial value, not only do you have to pull a volume of material out of it, but the process has to be repeatable. Your success cannot be 1 or 2 percent. A 2 percent success rate is not a success, it's a biological accident. Where's the other 98 percent? Show me them.[11]

Dominko was by then working for Advanced Cell Technology (ACT), in Worcester, Massachusetts. The chairman and CEO was Michael West, who had already founded Geron Corporation and had achieved "remarkable success as a kind of merchant of immortality."[12]

Since then, he has tried to clone a human being. But by 1999 he saw that stem cells were the coming thing, and was promoting the idea that "stem cells and related technologies might someday completely revise the tables of average human life span."

Today, according to ACT's website, West is "pursuing strategic collaborations in the state of California with members of academia, industry, and foundations to further accelerate the pace of our research." By then California voters had passed Proposition 71, and $3 billion in funding would be distributed, some to qualified businesses in the state.

Advanced Cell Technology plans to establish a research facility in California. If the investors wouldn't pony up, maybe the taxpayers would oblige.

Stem cell research grew out of cloning. In cloning, the nucleus of an adult cell is transferred to an egg from which the nucleus has been removed. An electric shock is applied, and the imported DNA will fertilize the egg. Then it starts dividing and forms what is called a blastocyst, a hollow ball of cells that appears a few days after fertilization. Within that ball, an inner group of cells forms, and these are the stem cells. They are elusive, appearing only briefly before they begin to differentiate and specialize.

A Book You're Not Supposed to Read

Brave New Worlds: Staying Human in the Genetic Future by Bryan Appleyard; New York: Viking, 1998.

In cloning, the embryo develops into a fetus (maybe) and then into a fully grown body (maybe). With stem cells, the process is interrupted. The stem cells are not permitted to become more and more specialized through further division. They are extracted, then "bathed in proteins," in one account, and by such means "directed" to specialize in any desired direction. To benefit diabetics, for example, stem cells would be coaxed in the direction of turning into the pancreatic cells that make insulin.

But how to "coax"? No one knows how to do it.

A normally growing embryo, if left to its own devices, "knows" how to develop into a body with trillions of specialized cells, each one in its proper place playing its proper role. Scientists, by contrast, don't know the formula that will "direct" stem cells in any particular direction. This is why stem cell therapy looks as though it will be even harder to achieve than the cloning of dogs.

Some scientists have tried to reassure us, and perhaps have convinced themselves, that the control of stem cell development can be easily achieved. "With stem cells," said Ronald McKay, a researcher at the NIH in Bethesda, Maryland, "you tickle them and they jump through hoops for you."[13]

Not so—unless you don't mind which hoops they jump through. We will encounter more of McKay's adventures as ringmaster in the next chapter.

Chapter 9

THE STEM CELL CHALLENGE
TO BIOENGINEERING

When the American Association for the Advancement of Science met in San Francisco in 2001, two scientists said they had used embryonic stem cells to reduce the symptoms of Parkinson's disease in rodents. They were Dr. Ronald McKay of the National Institutes of Health, for whom (as we saw) stem cells jump through hoops, and Dr. Ole Isacson of Harvard Medical School. "We can take the embryonic stem cells through a series of transitions until they become the dopamine cells," said McKay.

Certain brain cells normally produce a substance called dopamine, but sometimes, for reasons not known, they lose their function. The result is Parkinson's disease. Isacson said his lab had injected into the brain specific cells extracted from the embryo. A natural process in the brain then transformed them into dopamine producers. "The cells organize themselves to become very functional," Isacson said. "We see the cells behaving in a way to reverse the symptoms (of Parkinson's) in the mouse and rat."

The next day the Associated Press led off with this:

Scientists may be on the brink of curing Parkinson's disease using transplanted embryonic stem cells, but where and when that new treatment is tested in humans depends on unresolved political decisions.[1]

Guess what?

.˙. Stem cell research is legal; there is nothing to stop private investment. But stem cell advocates want government funding. The truth is, the chances of success are not good, and the smart money knows that.

.˙. Dream or nightmare? *Scientific American* in July 2005 reported: "Embryonic stem cells, unlike adult stem cells, cannot be used directly in therapy because they cause cancer."

After a few sentences, the reporter added: "McKay and Isacson said researchers are almost ready to test the technique in humans, but social and political issues must be resolved in the United States before that step can be taken in this country. At the same time, McKay said it may happen soon in Britain, France, or the Netherlands, as those countries are adopting policies to advance embryonic stem cell research."

A BBC reporter took an identical tack. "Scientists are closer to finding a cure for Parkinson's disease using special 'master cells' taken from embryos," Jonathan Amos reported. But, he continued, McKay and Isacson "may find that political and ethical obstacles will delay the treatment getting into clinical trials."[2]

The good news was circulated immediately. Charlotte Mancuso, a Parkinson's Action Network volunteer, posted on the Parkinson's Information Exchange Network:

> This is not just good news for a cure, it's good news for those of us who wish to remove the political stumbling blocks from researchers struggling to bring about a cure. The political message here dominates, and that is exactly what we need in the press, and all the media for that matter. Let's milk this one for everything it's worth—our lives![3]

Science itself wasn't all that different. In its initial article on the report from the San Francisco meeting, the magazine wrote:

> As the future of stem cell research teeters on the unsteady ground of political controversy, the case for its potential benefit grows more solid. At the AAAS annual meeting, neurotransplant researcher Ole Isacson of Harvard Medical School in Boston reported that embryonic stem cells implanted in the brain of rats and mice grow into the types of cell that wither in Parkinson's disease.[4]

The early reporting on the Parkinson's "cures" anticipated the story-line that would remain unchanged in the United States over the next four years. The scientists are making great progress, but they have been hamstrung by the politicians. Many Americans today are convinced that the main obstacle to cures lies in our politics, not in our limited understanding of cell biology.

Overwhelmingly, the stem cell debate has been a political debate. At times, the science could barely be heard above the din. If progress was slow in coming, someone had to be blamed. Readers looking for facts on stem cells—can they replace diseased tissue or not?—would find answers difficult to come by. Evidently, many scientists and reporters preferred the issue framed as one of theology versus science. Religious obstructionism, after all, is a more appealing adversary than a deficit of understanding. For it is beginning to look as though scientists have over-promised, a fault to which government-funded science is all too prone.

Six months after the San Francisco meeting, in August 2001, George Bush permitted federal money to be used for research on some embryonic stem cell lines—those that already existed. Until then, Congress had disallowed funds for all such research. At the time, Bush's decision was seen

Now I Feel Better

"Critics point to worrisome animal research showing that embryonic stem cells sometimes grow into tumors or morph into unwanted kinds of tissues—possibly forming, for example, dangerous bits of bone in those hearts they are supposedly repairing. But supporters respond that such problems are rare and a lot has recently been learned about how to prevent them."

Rick Weiss, "The Power to Divide," *National Geographic*, July 2005

as a partial victory for science. Later, that victory was quietly pocketed and Bush was criticized for retaining any restrictions at all.

But things didn't go as expected. Before long, less promising news about the treated rats and mice surfaced. When Isacson injected stem cells into the brains of rats, *Science* subsequently reported, one-fourth of the animals showed no effect, half showed some improvement, and the remaining one-fourth developed tumors and died.[5]

An even less encouraging story had already been given front-page coverage by the *New York Times*:

Parkinson's Research Is Set Back by Failure of Fetal Cell Transplants

The new study had revealed a "disastrous side effect." In about 15 percent of the (human) patients, the cells grew only too well, emitting so much dopamine that they "writhed and jerked uncontrollably." Worse, there was "no way to remove or deactivate the transplanted cells."[6]

To be sure, the study used cells from aborted fetuses, not stem cells. But the story, by Gina Kolata, was notably bleak. She had been burned three years earlier by DNA star Jim Watson, who had told her at a dinner party that Dr. Judah Folkman "is going to cure cancer in two years." That had appeared on the front page of the *New York Times*, unleashing successive waves of hope and disappointment.[7] Kolata has covered promised medical breakthroughs with caution ever since. Stem cell promises especially.

In fact, subsequent reporting on stem cells in most mainstream media has been wary, at least on the inside page. Cures are probably decades away, journalists often say. Optimism about treatment for Alzheimer's has almost disappeared, although Nancy Reagan and Patti Davis, who spoke about an Alzheimer's "cure" at the Democratic convention in 2004, remain true believers.

Most stem cell research has been conducted using rats and mice. Mouse stem cells were first isolated in 1981 by Martin Evans at the University of

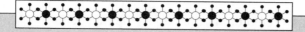

Dream or Nightmare?

Scientific American teamed up with the *Financial Times* for a special report on stem cells in the July 2005 issue. "Investigators are confident that someday stem cells will be the foundation for fantastic cures and therapies," said editors John Rennie and Lionel Barber.

The lead article, "Mother of All Cells," was by Clive Cookson, who reported: "Embryonic stem cells, unlike adult stem cells, cannot be used directly in therapy because they cause cancer. Indeed, one laboratory test for embryonic stem cells is to inject them into mice and analyze the teratoma (a tumor formed of fetal tissue) that arises."

Guiding the differentiation of these cells is "a scientific nightmare," he added.

"Mother of All Cells," *Scientific American*, July 2005

Cardiff, in Wales, and simultaneously by Gail Martin at the University of California–San Francisco. Martin also named them embryonic stem cells.

Caution is in order, because we cannot assume that what works in mice will work in men. Mice, for example, can be cured of diabetes by the transplant of pancreatic cells from other mice. In humans, the immune system must first be suppressed or it will wipe out the transplants, perceived as invaders. In the 1990s, when human stem cell lines were grown in the lab, they were grown alongside mouse cells in the petri dish. For unknown reasons, the mouse cells prevent the human stem cells from spontaneously "specializing." It's as though the stem cells "sense" they are in the wrong place and decide that the best course of action is to do nothing.

Human embryonic stem cells were first isolated and grown by two teams of scientists not that long ago, in 1998—one at the University of

Wisconsin and the other at the Johns Hopkins School of Medicine. The *Washington Post* reported at the time:

> The cells multiply tirelessly in laboratory dishes, offering a self-replenishing supply from which scientists hope to grow replacement tissues for people with various diseases, including bone marrow for cancer patients, neurons for people with Alzheimer's disease, and pancreatic cells for people with diabetes.[8]

Stem cells are removed from an embryo at an early stage and have the "potential," as we are always told, to turn into any cell of the body. They are "pluripotent." It is important to realize that this is true by definition.

A fertilized egg is "totipotent." We know that it can turn into any cell in the body because it does in fact multiply into every cell in the body. The confidence with which researchers assert that stem cells have the potential to become any cell is an expression of this tautology. What looks like an empirical claim is a logical necessity.

There are also adult stem cells, which seem to persist in the mature body as sources of replacement cells. Because research on adult cells does not involve the destruction of embryos, they are politically uncontroversial. Furthermore, they do not arouse the body's immune system. So they seem promising. And great medical claims have been made on their behalf. But the evidence is anecdotal, and not reliable. There are glowing reports, certainly. We hear of recoveries from paralysis after spinal-cord injuries and heart-muscle tissue

Paper Profits

As to the ability of adult stem cells to revert to other types of tissue, Dr. Irving Weissman told the *New Yorker*:

"I wanted that to be true. I mean, first, we actually did discover blood-forming stem cells, so I feel a little bit of the glory would rub off on me. Also I have a company that has the patent, and a license from Stanford—I could be the richest person on earth! I had every possible motive to want that to be true. But it's not. Every one of the claims we tracked down turned out not to be true."[9]

regenerated with adult stem cells from bone marrow. In a Texas Heart Institute trial, one doctor said, some bedridden patients, after stem cell treatment, "were "jogging on the beach, one climbed eight flights of stairs, and, one who had gone home to live with his mother, reopened his business."

Clinicians are encouraged, but researchers are more skeptical. The *New York Times* reported:

> At least two separate laboratories, at Stanford University and at the University of Washington in Seattle, reported last year that they had been unable to repeat the Orlic-Anversa [heart-tissue] experiment. Bone marrow stem cells, these researchers found, did not turn into heart tissue. The few that lodged in the heart turned into blood cells in the usual way. The Stanford researchers, who included Dr. Irving Weissman, a leading expert on the blood's stem cells, warned that until the science underlying the clinical trials was better understood, "these studies are premature and may in fact place a group of sick patients at risk."[10]

Before the 2004 election, Weissman was interviewed for an article on California's Proposition 71, an initiative making $3 billion available for stem cell research. He was the first scientist to identify and isolate adult stem cells in any species, and the first to isolate blood-forming stem cells in humans.

"Just identifying a true stem cell can be tricky," two researchers wrote in *Scientific American*. "For all the intensive scrutiny of stem cells, they cannot be distinguished by appearance. They are defined by their behavior." But that behavior itself is precisely what is at issue, and what scientists are trying to control . . . with very limited success. It was a remarkable admission by the authors, who have considerable expertise in the field. In fact, it would seem to cast a pall of doubt over the whole field.[11]

The University of Wisconsin team, led by James A. Thomson, used "leftover embryos" from fertility clinics, donated by parents, while the

Johns Hopkins team was led by John D. Gearhart, who retrieved his stem cells from the developing gonads of aborted fetuses. The federal funding of human embryo research had been banned by Congress four years earlier, so neither group used any federal money. In Thomson's case, "not even an electrical extension cord had been bought with federal funds," the *Washington Post* reported.[12]

With research looking promising, commercial rights were licensed to Geron Corporation, the California company that funded the research.

By 1999, the science looked golden. About a year after the news about embryonic stem cells was publicized in many a front-page story, *Science* published one of its "Breakthrough of the Year" articles with the headline: "Capturing the Promise of Youth." It referred to the "extraordinary potential of stem cells," perhaps "to heal many kinds of illness." The year had been one in which

> development biologists and biomedical researchers published more than a dozen landmark papers on the remarkable abilities of these so-called stem cells. We salute this work, which raises hopes of dazzling medical applications and also forces scientists to reconsider fundamental ideas about how cells grow up, as 1999's Breakthrough of the Year.[13]

Stem cells may be used to treat diseases in all sorts of ways, "from repairing damaged nerves, to growing new hearts and livers in the laboratory; enthusiasts envision a whole catalogue of replacement parts," the magazine believed. It had been "a turning point." Both science and society had recognized our "newfound ability to manipulate a cell's destiny."

That was misleading, coming from the nation's leading science journal, because it sounded as though we now really could "manipulate a cell's destiny." But that was still far from being the case. It implied that the goal of the entire exercise had already been achieved.

Sixteen months later, however, the *New York Times* reported on its front page something that did appear to be a real breakthrough:

> Scientists Report 2 Major Advances in Stem-Cell Work; Debate
> Likely to Heat Up Advances Come as Bush Weighs Argument
> of Researchers and Foes of Abortion

In the first of the two advances, Nicholas Wade reported:

> Biologists at the National Institutes of Health used mouse
> embryonic cells to generate insulin-producing organs resem-
> bling the islets of the pancreas, a feat that holds promise for
> treating Type 1 diabetes, also known as juvenile diabetes.[14]

The work had been conducted at the NIH by our old friend Dr. McKay, still working with mice, but now focusing on diabetes. His lab had apparently found a way to coax embryonic stem cells down the path from pluripotency to the pancreatic precursor cell. A "five-step method" had induced mouse stem cells to morph into the insulin-producing cells in the pancreas.

They produced not only the insulin that tells cells to take up glucose from the blood, but also three lesser-known hormones. Embryonic cells really had grown into these differentiated cells, McKay and his cowork-ers claimed. The new cells had assembled themselves in the petri dish, even secreting insulin when exposed to glucose.

"This seems to be the first time that a miniature organ has been coaxed to form from embryonic stem cells," Nicholas Wade

"Get My Broker on the Phone"

Drug developer Geron's stock price, nearly $70 in March 2000, fell to $1.70 by 2003, suggesting that investors were less enthusiastic about stem cells than some scientists. More than half the company's employees were laid off. But after California voters said yes to Proposition 71, things looked up. Geron is expected to be a beneficiary. If so, tax-payers will substitute for investors; the company's stock price today is around $10.

reported, "proof that in the right circumstances the body is a self-assembling system." The *Times* devoted its entire "Science Times" section to stem cells on December 18, 2001. Discussing the new discovery, Wade concluded:

> Embryonic stem cells possess and can gain access to the entire manual of genetic instructions for generating and regenerating the body. Given the bare minimum of appropriate cues, it seems, they will mold themselves into the components of the right tissue. Human embryonic stem cells are expected to behave in the same general way as mouse cells, although few studies of them have been done.

Other researchers followed suit, claiming they had turned human stem cells into islet cells. This was good news indeed, for one of the main problem facing diabetes patients hoping for islet-cell transplants is the shortage of donors. The discovery gave new hope, the *Times* reported. Human embryonic cells "can in principle provide an inexhaustible source of islets, and of many other critical tissues that are damaged in disease."

In July 2004, Dr. Robert Goldstein, director of research for the Juvenile Diabetes Research Foundation, testified before Congress that "recent studies have demonstrated the ability to coax embryonic stem cells into insulin-producing cells in the lab."

But research by Dr. Douglas Melton of Harvard has cast doubt on all these findings. Melton himself has two teenage sons

On the Never-Never

"With some conditions, such as Alzheimer's disease, science has yet to understand what goes wrong. Simply replacing damaged brain cells with new ones grown in the lab from stem cells isn't yet feasible and may not be for decades, researchers say."

The *Wall Street Journal*,
August 12, 2004

with juvenile diabetes, and has dedicated his career to finding a cure. The disease afflicts about one million Americans and is treatable by the self-injection of insulin, but islet cells do a much better job. Often, in the end, patients face amputations and blindness. Melton remains convinced that stem cells can be turned into insulin-making cells, but admits, "We just don't know how."

Less than a month after Goldstein's testimony, the *Wall Street Journal* published these details of Melton's findings in a front-page story.

> Biggest Struggles in Stem-Cell Fight May Be in the Lab
>
> Dr. Melton says that Dr. Goldstein and the researchers who reported the early results are wrong. In published reports and at numerous science meetings over the past two years, Dr. Melton has criticized the findings. He says that in their attempts to grow islets, the research teams used chemical growth factors, including insulin. Dr. Melton says the insulin in the mix confused the results.[15]

Robert Lanza and Nadia Rosenthal said the same thing in *Scientific American*. McKay's stem cells "had absorbed insulin from their culture medium rather than producing it themselves," they wrote in June 2004. And now Ronald McKay was telling the *Wall Street Journal*: "Anyone who says that new therapy is around the corner or even a few years away is just wrong."

Melton, incidentally, had followed the Bush administration's guidelines by pursuing his research without federal money. His support came from the Howard Hughes Medical Institute, Harvard University, private sources, and, ironically, the Juvenile Diabetes Research Foundation whose research director he had contradicted.

The body's immune system is an obstacle facing all embryonic stem cell research. This is particularly true of diabetes, in which a hyperactive immune system is responsible for the illness in the first place. It attacks and destroys the insulin-producing cells, mistaking them for outsiders. But cells deriving from an embryo really are outsiders, and the immune system may be expected to return to its task with a vengeance.

It was only belatedly that this surfaced as an issue. *Science* addressed it in an article published in 2002. It had been found, "contrary to some researchers' early hopes," that the immune system was as attentive to embryonic stem cells as to any other foreign bodies, so that they could not be smuggled in. The growing cells, as they became more differentiated, express increasing levels of the "labels" that the body uses to distinguish between what is native and what is foreign.

Therefore, a patient's immune system is likely to reject transplanted tissue derived from embryonic stem cells. "Scientists hoping to use the cells to treat Parkinson's disease, diabetes, and other maladies will therefore have to find ways to reconcile the body's defense system with the transplanted cells," *Science* reported. Previous interpretations had been too optimistic, and had raised false expectations—the oft-repeated story of biotechnology in a nutshell.[16]

In early summer 2005 there was a renewed burst of enthusiasm. Researchers in South Korea had used a technique called somatic cell nuclear transfer to derive human stem cell lines that are genetically matched to patients. This could solve the problem of rejection. The nucleus of a donated egg is removed and replaced by that of a cell from the patient. This fertilizes the egg and the patient is then cloned. After the early embryo has grown to about one hundred cells, it is dismantled to harvest its stem cells. As these stem cells are intended for the medical benefit of the patient who donated the somatic cell, the technique is called therapeutic cloning.

It is impossible to say at this stage whether the new technique will defeat the immune system. As always, theory will have to yield to practice. Meanwhile, government-funded research continues without restrictions not just in South Korea, but also in Singapore, Japan, Sweden, Israel, and Britain.

Private investors have all along been free to put up their own money, and among those who opened their wallets to support California's Proposition 71 were Microsoft's Bill Gates, the founders of eBay and Amgen (a profitable biotech company), and assorted Hollywood producers. Even if Congress does not fall into line, other states are sure to fund research, if only discourage researchers from moving to California (as Michael West seems to be doing with Advanced Cell Technology).

The time will come, if we are not there already, when money will no longer be a plausible excuse and scientists will have to acknowledge the limits of their own understanding as they grapple with the immense complexity of the cell. What nineteenth-century researcher Theodor Schwann had called a "cavity" filled with a "homogenous transparent liquid" has turned out to be more complex than we can even imagine. Ever since Darwin, the tendency has been for scientists to regard biological systems as much simpler than they really are. (That way, it is not so difficult to believe that they were assembled over millions of years by trial and error.)

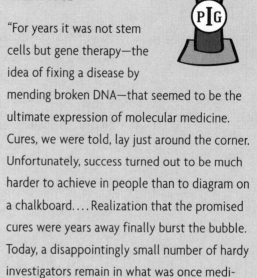

VERITAS

"For years it was not stem cells but gene therapy—the idea of fixing a disease by mending broken DNA—that seemed to be the ultimate expression of molecular medicine. Cures, we were told, lay just around the corner. Unfortunately, success turned out to be much harder to achieve in people than to diagram on a chalkboard....Realization that the promised cures were years away finally burst the bubble. Today, a disappointingly small number of hardy investigators remain in what was once medicine's most highly anticipated new area of research.

"Is this going to be the fate of embryonic stem cell science in five or ten years? I hope not, and yet it's also not very difficult to imagine this happening. Already newspapers are filled with extravagant claims of progress and cures. These reports belie the very slow rate of true scientific advancement. Add to this the explicit expectation of rapid clinical progress... and you have a recipe for trouble."

David A. Shaywitz,
Harvard stem cell researcher, *Washington Post*, April 29, 2005

The overriding problem is something that has perplexed embryologists for over one hundred years and continues to overwhelm researchers to this day. They do not know how the body is assembled into an organism of 100 trillion cells, starting with a single fertilized egg, when all the cells contain the same DNA—the same set of genetic instructions. Why is the body a chiseled multi-functional marvel, and not just a great meatball of cells?

We do not know. Geneticists do know that certain genes are switched "on" and "off" at different times, but, for all their talk of "decoding" the genome, we do not know how or why this happens. It remains a mystery. Our stem cell maestros talk about "coaxing cells" and "five-step processes," but those are little more than polite names for guesswork. Sometimes, it's true, desired things do seem to happen, but when they do it seems to be fortuitous and not easily repeated. That has been the cloning story in a nutshell.

The assumption that cells can be removed from their natural environment and "coaxed" in the lab dish in various desired directions is an expression of the hubris of science. It is comparable to believing that education could be improved nationwide if children were taken from the home environment, or removed from high school, on the grounds that such traditional methods of schooling do not reliably prevent truancy, juvenile delinqency, or even occasional Columbine high-school tragedies. So, capture them at birth and coax them in desired directions with special nutrients and tutoring!

The problem is that the scientists don't know have any idea what that special training should be. Stem cell science has produced some very ugly things called teratomas—tangled meatballs of teeth and hair growing and extruding where stem cells were injected. Maybe in the end the home environment will be shown to have worked the best, in both the social and the cellular environments, even if some cells, and some students, do quit or rebel or become "anti-social elements" from time to time.

Genetic engineering is turning out to be as hard to achieve in our day as social engineering was in the Communist era. Curiously, diametrically opposed errors seem to be involved. Social engineering was derided as unattainable and failed because "nature" was overlooked. Genes were disregarded as unimportant. Politicized scientists such as Trofim Lysenko [1898–1976] obeyed their political masters and claimed that seeds, when properly "trained," could yield a bountiful crop in Siberian weather. It was all faked for the benefit of Joe Stalin.

The engineering of the cell, in contrast, has downplayed the role of "nurture," or the natural environment of the cell. Genes are now thought to be autonomous and all-important—wrong again.

In an article for the *New York Times*, Harvard zoologist Stephen Jay Gould discussed stem cell research with reference to Von Baer's law. Resident in Germany and Russia in the nineteenth century, Von Baer formulated the "central principle of embryological development." The growing organism becomes ever more differentiated. There is "no turning back after the blueprint becomes finalized from a broad mass of initial potential." The law "gives us no alternative to embryonic stem cells for now," Gould wrote. It may be that he is right, and that adult stem cells can't go back on the course they have taken.

But Gould's real aim in making this point was to repeat a famous remark that Von Baer had made in his old age. He said that all new and important ideas "must pass through three stages: first dismissed as nonsense, then rejected as against religion, and finally acknowledged as true."[17]

The case of stem cells suggests that new ideas today are likely to enjoy a rather different reception: first hailed as true, then bolstered by religious opposition, and finally acknowledged as false.

Chapter 10

A MAP TO NOWHERE

Some of the controversies in this book may become conventional wisdom—the need to bring back DDT, for example. Others seem headed that way. Nuclear power is rapidly becoming a "green" cause. Other issues are a matter of heated debate. Global warming is the best example; stem cells are another. Almost overnight, the adequacy of evolutionary theory seems to have become a front-page story.

Still other stories have been flying beneath the radar. Hardly anyone says a word, and no one is making a fuss. But there are surprises in store. A striking example is the Human Genome Project, which is not working out as expected.

Genetic engineering began to seem achievable in the 1970s, when it was shown that "phage" viruses could slice the DNA of bacteria. This led to new ways of recombining the genetic material. Segments of DNA, known as genes, are sometimes mutated, and then they don't work properly. These dysfunctional segments could now be replaced by new segments that worked properly. That was the theory, at any rate.

Before you knew it, a distant possibility became "a new era of medicine." Newspapers overflowed with discussions of the new breakthroughs achievable with "recombinant DNA." At the Asilomar Conference in 1975, a group of influential scientists drew up voluntary guidelines to ensure the safety of recombinant DNA. The Human Genome Project was soon

Guess what?

.ᐧ. Investors and tax-payers have spent millions—and perhaps billions—of dollars on genetic engineering, without success.

.ᐧ. When the Human Genome Project was completed, thousands of genes already patented by private companies were shown to be nonexistent.

.ᐧ. The genetic defects that cause hereditary diseases such as cystic fibrosis were known long before the Human Genome Project started.

cranked up and enthusiasm knew no bounds. Medicine as we knew it was about to be transformed.

A leading preacher of the new gospel was W. French Anderson of the University of Southern California. The new genetic technology will "revolutionize medicine," he wrote in *Scientific American*. There had been three previous revolutions. First sanitation and public health, then surgery with anesthesia, followed by vaccination and antibiotics. Gene therapy would be the fourth. The delivery of selected genes "can potentially cure or ease the vast majority of disorders," Anderson wrote. "Almost every illness arises because one or more genes are not functioning properly."[1]

There was no warrant for that claim about the genetic basis of "almost every illness," but hundreds of researchers were saying much the same thing. A consensus, as we have seen with the clamor for stem cell research, has a way of becoming irresistible, and those who espouse consensus in science do so for just that reason.

Now, however, some are beginning to wonder if gene therapy will ever be a reality. For after thirty years of experimentation, it has had almost no successes. (I will return to the exception.) The oft-promised revolution of medicine turns out to resemble a rainbow: it's always over the horizon—or twenty years in the future.

"Within twenty years," Anderson predicted (in 1995), gene therapy "will be used regularly to ameliorate—and even cure—many ailments." The director of the National Human Genome Research Institute, Francis S. Collins, has predicted that by 2010 "there will be preventive treatments for a dozen common genetic illnesses." It hasn't happened yet, but the budget of his taxpayer-funded institute has soared, from $27 million in 1988 to $500 million today.

In June 2000, J. Craig Venter of Celera Genomics and Francis S. Collins were introduced by President Clinton at the White House. The headlines were across all the front pages: scientists had assembled a "working draft" of the "entire human genetic code." It was compared to

Newton's discoveries and the 1969 moon landing. "A New Era Begins," was just one headline in the *New York Times*. The iconic double helix of DNA was everywhere.

The announcement "marked the end of an arduous, ten-year, $2 billion international effort to identify and place in order all 3.1 billion molecular 'letters' of DNA residing inside virtually every human cell," the *Washington Post* reported.[2]

Genome is decoded—again

The end? Venter and Collins were back again the following February with a more complete genome. Two hundred and fifty journalists were packed into a Washington hotel ballroom, and James Watson of DNA fame took a bow along with Senator Pete Domenici of New Mexico. Along with Francis Crick, Watson had discovered the "twisted ladder" structure of DNA, the long molecule that is the site of all the genes. The Venter and Collins findings were soon to be published, with a comical abundance of co-authors, in the journals *Nature* and *Science*.

In his introductory remarks to the media assembled, Senator Domenici said at the podium that James Watson had just whispered to him: "'You must say that this project was congressionally driven.' And that's true," the senator added. "This project, in terms of the U.S. government, was truly started in the Congress."

There was one big news item. Earlier, the number of human genes had been estimated at perhaps 100,000 or even 150,000. *Scientific American* had published a story beginning: "Now that all the 100,000 or so genes that make up the human genome have been deciphered." But Venter and Collins had a different count: Maybe 30,000 or 35,000 genes? We have "only twice as many genes as a fruit fly, or a lowly nematode worm," said Eric Lander, head of genome research at the NIH-funded Whitehead Institute in Cambridge, Massachusetts. "What a comedown!"

The journalists roared on cue, and that would be the sound bite for National Public Radio.

There was another implication, however, and it has still hardly begun to sink in. The way genes work "must be far more complicated than the mechanism long taught." said the *Washington Post,* almost in a whisper, as though discussing a death in the family. If the new gene number was correct, the genetics textbooks would have to be rewritten. And the unmentionable thought came unbidden: the therapeutic breakthroughs were surely going to be much harder to achieve than anyone had realized. Because they really hadn't figured out this stuff at all.

Craig Venter's opening statement contained the bombshell. Since the meeting at the White House only eight months earlier, he said,

> our understanding of the human genome has changed in the most fundamental ways. The small number of genes—some 30,000—supports the notion that we are not hard wired. We now know the notion that one gene leads to one protein, and perhaps one disease, is false.
>
> One gene leads to many different protein products that can change dramatically once they are produced. We know that

Now They Tell Us: Genetic Checkers

"When people talk about genetic engineering today, it's really kind of a joke because they mean, 'I moved a gene from one organism to another organism and I'm going to pray that it works.'"

Michael Ellison, a biologist at the University of Alberta, *New York Times* "Science" section, August 16, 2005 some of the regions that are not genes may be some of the keys to the complexity that we see in ourselves. We now know that the environment acting on our biological steps may be as important in making us what we are as our genetic code.[3]

The old dogma, embedded in the textbooks for sixty years, and prevailing to this day, was that one gene made one protein. George Beadle and Edward Tatum had won the 1958 Nobel Prize for formulating this doctrine. Usually it is said: "one gene, one enzyme," but an enzyme is a special kind of protein, so it comes to the same thing. Now, in front of some of the country's most eminent molecular biologists and the whole media tribe, Venter was disclosing that there may be *ten times* as many proteins as genes. For we have "perhaps 300,000 proteins," he said.

The genome consists of a string of four nucleotide bases, symbolized by the letters A, C, G, and T, and there are 3.1 billion of them. Over 98 percent of this immense string, seeming to have no role, was for years dismissed as "junk DNA." Its stray and irrelevant blocks of gibberish, millions of letters long, were doing nothing in particular, geneticists reassured one another. It was mere rubbish accumulated during our evolutionary history and was construed as evidence for evolution. (For why would a designer God be so untidy as to leave all that junk lying around?)

Neither Collins nor Venter accepted the "junk" argument, however. Both said we can't just assume it has no role because we don't know what it is. More recently, the argument that most of the DNA is "junk" has itself been junked. Articles in *Science* in 2003 allowed that researchers were finding "gold in the junk,"[4] and that has been a bonus from the genome project. But with all those additional bases now in play, the task confronting the genetic engineers has grown at least a hundred times more complicated.

The intermittent "coding" segments along the way are called genes, and they give instructions for the manufacture of the body's proteins. In people with heritable diseases, some of these proteins are defective (or absent). So the new genomics would find the defective genes along the DNA string, and with the pristine version substituted, the protein it made would also be restored. That was the theory.

Now they were telling us that genes are not just simple "strings"—contiguous segments of DNA, each one coding for one message—but combinations of separate segments strung along the genome. Between them lie intervening segments that can be cut out, or not cut out, or cut out some of the time and not at other times. And the relevant or coding parts (called exons, as opposed to the intervening parts, called introns), can be put together in numerous different ways to send different messages and make a variety of proteins as the occasion demands.

At the press conference, Francis Collins was asked if the smaller number of genes would make medical advances easier or more difficult. "I would say easier," he said. Every gene-search is like trying to find a needle in a haystack. "Guess what? The haystack just got three times smaller."

He looked a little sheepish as he said it, and in response to a similar question about the combinatorial interaction, Collins retreated from the haystack metaphor. Craig Venter said more simply that when you consider there is maybe a "tenfold expansion" in the number of proteins compared to the number of genes, it "does indicate increased complexity."

It surely does.

Imagine that an intelligence service discovers coded messages sent by a spy. At first they assume that the task of decoding will be straightforward. But on closer inspection, the message means one thing if the signal has been received and acted upon, another thing if it has been received and not acted upon, another thing if the receiving apparatus is not switched on, and so on. Rather than a code, we have something more like a set of rules for a complex interactive game. There are feedback loops, and circuits within circuits, and a lot of things happening inside the cell but *outside* the genome, in the unfashionable realm of cytogenetics. NIH-funded geneticists didn't even want to think about that, because they had imagined that by sticking to the four nucleotide bases, they had the problem neatly "digitized." Computers would hum away unaided, twenty-four hours a day, and unravel the mysteries for them while they slept.

The new gene number gave rise to much embarrassment. At the time of the White House gathering, David Baltimore, the president of Caltech and the winner of the 1975 Nobel Prize for medicine (for work in molecular biology), had written an article for the *New York Times*. It was headlined "50,000 Genes and We Know Them All (Almost)."[5] Actually, they barely understood the genome at all at that point. But in generating a mood of consensus, every bit of Nobel certitude helped. "Humans have no more genetic secrets; our genes are a book open to all to read," Baltimore wrote, erroneously. Eight months later, in the issue of *Nature* announcing the new analysis of the human genome, he wrote (more soberly):

> We wait with bated breath to see the chimpanzee genome. But knowing now how few genes humans have, I wonder if we will learn much about the origins of speech, the elaboration of the frontal lobes and the opposable thumb, the advent of upright posture, or the sources of abstract reasoning ability, from a simple genomic comparison of human and chimp. It seems likely that these features and abilities have mainly come from subtle changes . . . that are not now easily visible to our computers, and will require much more experimental study to tease out. Another half century of work by armies of biologists may be needed before this key step of evolution is fully elucidated.[6]

Not in the genes? We always knew that . . .

The old dream of reducing biology to physics, of believing that something simple—four nucleotides on a string—could explain the vast complexity of the human (or any other) body, had received a serious blow. Many of the biotech companies had been misled by the idea that computing power

Read the Headlines

**Genetic Code of Human Life
Is Cracked by Scientists**
New York Times, June 27, 2000
[front page]

The Genome Is Decoded. Be Happy.
Wall Street Journal, February 14, 2001

**The Genome Is Mapped.
Now He Wants Profit.**
New York Times, February 24, 2002

**Dream Unmet 50 Years
After DNA Milestone;
Gene Therapy Debacle
Casts Pall on Field**
Washington Post, February 28, 2003
[front page]

**Once Again, Scientists Say
Human Genome Is Complete**
New York Times, April 15, 2003

was all it would take. It was an error comparable to that of Ernst Haeckel in nineteenth-century Germany, who dismissed the cell as "a simple little lump" of protoplasm. Now we know it to be a minute factory of daunting complexity, no more resembling a simple lump than a semiconductor fabrication plant resembles a child's sandcastle on the beach.

Within months of the press conference, the word "genomics" began to fade away, to be replaced by "proteomics." The "proteome" is a far more complex thing, because proteins are not just one-dimensional strings of data, but complex sets of amino acids folded compactly into three dimensions. And no one knows how they get the instructions to do that.

Francis Crick was invited to the Washington hotel event, but sent a videotape instead. "We foresaw very little of what happened in molecular biology," he said. The latest development would have an "enormous impact on medicine." But he seemed concerned. He hoped that on balance the news would bring "more good than evil."

Afterwards, a crowd clustered around James Watson, an aged star still on stage and enjoying it. Forty-eight years had passed since his joint publication with Crick of the double helix. "Can one gene really generate ten proteins?" someone asked him.

"Some genes can give rise to fifty different proteins," he said.

No problem! He was unruffled, still in control. His demeanor indicated that the new knowledge about genes and proteins would be smoothly integrated into the received wisdom of molecular biology. The higher councils were taking it all in stride. But those who wanted to put it to medical use were beginning to furrow their brows.

The journalists never did quite get the story. There had been a "race" between Venter (private sector) and Collins (government) to complete the genome, and that was the way it was covered. (The rivals stage-managed a tie, hence their joint appearances.) By making rivalry the top issue, the journalists implicitly endorsed the underlying claim: that the rivals were racing to complete a task that really would revolutionize medicine.

Right now, it's not looking so good.

Worthless genes

There was another embarrassment, but on a different plane. Even after the press conference, Dr. William Haseltine of Human Genome Sciences assured the *Boston Globe* that there must be about 120,000 genes. Must be? The world's best genetic minds had missed tens of thousands of them in their haste to produce the completed map, he said. Maybe they had missed "as many as two-thirds of the genes that exist." Venter, Collins, & Co. were "guilty of sloppy science and sloppy conclusions." Their picture had "gaps, big gaps."[7]

Haseltine, who had studied under Nobel Prize winner Walter Gilbert at Harvard and joined the Harvard Medical School faculty, had become rich, worth more than $300 million in stocks and options after founding Human Genome Sciences in 1992. It was left to Andrew Pollack of the *New York Times* to summarize the key point: genes already patented were now shown to be nonexistent:

> Incyte Genomics advertises access to 120,000 human genes,
> including 60,000 not available from any other source. Human

Genome Sciences says it has identified 100,000 human genes, and Double Twist 65,000 to 105,000. Affymetrix sells DNA-analysis chips containing 60,000 genes.[8]

Those patents "could be worth less," Pollack added—or worthless.

Craig Venter told the London *Observer* that the head of a biotech company had phoned him in distress because he had already done a deal with the pharmaceutical company SmithKline Beecham to sell them details of 100,000 genes. "Where am I going to get the rest?" the man asked.

Isolated voices were heard to say that the gene mania of late twentieth-century America should be compared to the tulip mania of seventeenth-century Holland. But most people kept quiet. Journalists didn't want to lose access to their sources, scientists didn't want to lose their funding, investors didn't want to lose their investments. But many of them did just that. The principals were doing just fine, however, and the band played on. At the time of the fiftieth anniversary of the double helix discovery, the Whitehead Institute's Eric Lander said that Watson and Crick had discovered "the secret of life." And with the human genome now decoded (again), he added: "We have before us the instruction book of medicine."[9]

Not yet, alas.

Lander showed that the pride that scientists imputed to the human race was still much in evidence—from scientists themselves. Was it really so unexpected that the "blueprint" for making something as complex as a human being should itself turn out to be extraordinarily complex? Imagining otherwise was typical of the hubris of modern science.

In fact, the scientific obstacles to the predicted revolution had been daunting from the beginning. And unlike the gene count, these problems were once well understood by the leaders in the field. But many of them were beginning to believe their own press clippings.

The unsolved problem

Mutations in the genome arise at two levels. There are, first, *germ-line* mutations. They exist in the embryo, right after the egg is fertilized, and as the initial cells divide and keep on dividing until the body is fully grown, the initial mutation is cloned (copied) in all the cells of the body. (No one knows how many cells there are, but some say a hundred trillion, some say three times that! Imagine ten thousand times the number of human beings alive on earth today, and you have a low estimate of the number of cells in one human body.)

In the grown body, cells continue to divide, although at a rate that depends on location. This cell division, called mitosis, occurs frequently in the skin and in the gut, rarely in muscles, never (or almost never) in brain cells. Here, too, mutations may arise, perhaps accidentally, or for various other reasons: ultra-violet light or radiation might strike a cell, for example, damaging the DNA. These are called *somatic* mutations, and they arise only in the small number of cells affected, plus their "daughter" cells (if any).

The true genetic diseases—the ones we have all heard of, such as sickle-cell anemia, cystic fibrosis, and muscular dystrophy—are caused by germ-line mutations, which by accident have arisen in copies of a particular gene that the infant inherited from both mother and father. The mutation then shows up in every cell in the body, giving rise to disease. So these diseases are indeed caused by a defect in the genome. They may be compared to those rare times when a single typographical error alters the meaning of a text. (But most "typos" are immediately apparent as such; they cause embarrassment to proofreaders but no impediment to understanding.)

The heritable nature of these diseases was known before the Human Genome Project, and it was known by tracing their history in particular families. In some cases, the nature and exact position (on the chromosomes) of the mutation was also known—even before the genome project.

For example, the genetic defect giving rise to sickle-cell anemia was known fifty years ago.

"We've had our gene since 1989," Dr. Robert Beall, president of the Cystic Fibrosis Foundation, told the *Wall Street Journal* in 1999. Two years later the paper reported on its front page:

> Unsolved Mystery
> Cystic Fibrosis Gave Up Its Gene 12 Years Ago;
> So Where's the Cure?[10]

"Medical sleuths gain clues to workings of disease," the headline continued, "but most are dead ends." Yet gene therapy for the 10,000 cystic fibrosis sufferers in the United States has not been developed. In other words, the "breakthrough" with respect to mapping the human genome happened sixteen years ago in the case of cystic fibrosis, to no effect. Sickle-cell anemia can still only be treated by non-genetic therapy. And it is the same with all the other heritable diseases.

Recall that the genetic defect occurs in every cell. The great problem for genetic engineering is to put the "corrected" gene into enough cells to make a difference. It is still unsolved. There are hundreds of these genetic diseases, but they are rare, appearing overall in only about 1 percent of all births. In the serious cases, such as cystic fibrosis, they are rare almost by definition, because sufferers usually die before having children.

Therefore, it is said, these diseases "do not fit the business model." No money can be made by curing them. Nonetheless, hundreds of millions of dollars have been spent by philanthropic institutions to locate the genetic defects that cause them. Still, not much has come of these findings, beyond the patenting of screening tests, which can be used to warn couples, if both have the defect recessively, that any child they bear might well be born with the disease.

But biotechnology can expect little payoff from diseases that affect only a few thousand people. As a result, some years back, the focus of gene therapy quietly shifted toward far more common and potentially profitable diseases: cancer, heart disease, Parkinson's, and Alzheimer's. Collins's own lab at the NIH has engaged in a "huge and very complicated" search for genes causing adult-onset (type II) diabetes, and this soon became the new cause celebre for gene hunters.

But the idea that these conditions, in the aggregate affecting almost the entire human race, are caused by the sort of clear, isolated genetic "misspellings" that do explain sickle-cell or cystic fibrosis was entirely speculative. Today it seems to be just plain wrong. Cancer is the most important case, but I shall defer consideration of that explosive topic until the next chapter.

Consider now just those diseases that are truly genetic. Assume we know exactly what that error is, and let us also charitably assume that the knowledge came from the Human Genome Project. Why are we still unable to cure these diseases?

Gene therapy hits kids with SCID

Genetic engineering has had "almost" no successes, I said earlier. The single exception answers our question. In April 2000, it was reported that two infants born with a life-threatening immune-system disorder had received successful gene therapy in a French hospital. They had been kept inside a protective sterile "bubble," but a year later both were living normal lives. The *Washington Post*'s headline read: "Genetic Therapy Apparently Cures 2. French Team's Feat Would Be a First.[11] The disease is called SCID, an acronym for severe combined immunodeficiency. A single defective gene disables the T-cells that are the key to the body's ability to resist infections.

Remember, the task is to insert the corrected gene into enough cells to make a difference. Obviously it can't be done one cell at a time, and in all such cases the preferred agent—called a "vector"—is an immobilized

What Next?

"The reaction to the discovery that human beings do not have much more genomic information than plants and worms has been to call for a new and even more grandiose project. It is now agreed among molecular biologists that the genome was not really the right target and that we now need to study the 'proteome,' the complete set of all the proteins manufactured by an organism. Surely the very complex human being must have many more different proteins than a small flowering plant. Although the devotees of the genome project kept assuring us that genes made proteins and therefore when we had all the genes we would know all the proteins, they now say that, of course, they knew all along that genes don't make proteins."

Richard Lewontin, "After the Genome, What Then?" *New York Review of Books*, July 19, 2001

or otherwise harmless virus. Viruses do their thing by infecting as many cells as possible, but now they would be "transfecting" something good into the body: the crucial but absent gene.

Normally, an invading virus, whether it is harmful or not, is perceived as an "outsider" and runs into a serious obstacle—the body's immune system. But that was exactly what the infants with SCID lacked; they had to be sheltered inside a sterile bubble for that reason. And that made them the perfect target for gene therapy. But those with different diseases, such as cystic fibrosis, in which the immune system remains intact, are not so lucky. The defense department will still wipe out invaders, even if they are on a rescue mission.

The *New York Times* reported the development in Paris as a "dramatic breakthrough in a field that desperately needs one." With some asperity, the newspaper's editorial added: "The French success comes after a decade of hope, hype and recklessness in which gene therapists repeatedly implied that they were on the verge of delivering medical miracles only to have their attempts end in failure or death."[12] (Jesse Gelsinger, a teenager undergoing gene therapy at the University of Pennsylvania, had died in 1999.)

Sadly, in France, this optimism was misplaced. Three years later, in an article headlined "Dream Unmet After 50 Years," the *Washington Post*'s Rick Weiss reported the follow-up in Paris. The treatment, he wrote,

> has triggered a life-threatening form of leukemia, in which their renovated white blood cells are multiplying out of control and the boys are having to undergo chemotherapy to kill the very cells they had so desperately needed. The problem has brought dozens of gene therapy studies to a halt and has cast a pall over the struggling research field, which seeks to cure diseases by giving people new genes.
>
> In an awkward coincidence, the unfolding debacle will be the highlight of a Food and Drug Administration meeting today, fifty years to the day after James Watson and Francis Crick launched the modern age of genetic medicine by deducing, with the help of a cardboard model, the three-dimensional structure of DNA.[13]

The mouse genome was sequenced, the rat genome, the fruit fly genome, the chimpanzee genome, the puffer fish, the roundworm, baker's yeast. But not much came of it. The computers just kept churning out their mindless data. The human gene total fell below 30,000. By 2003 we had reached "mustard weed levels" and were "flirting with fruit fly territory," *Wall Street Journal* columnist Sharon Begley wrote.[14] Clearly, quantity (25,000 genes is the latest box score) is irrelevant. It is comparable to believing that Chinese is the most sophisticated language because its pictograms far outnumber our twenty-six-letter alphabet.

Venter hits jackpot, and Polynesia

The code is still not understood, despite the headlines. "It would take decades, or even centuries to completely understand the language of the code," James Shreeve wrote toward the end of *The Genome War: How Craig Venter Tried to Capture the Code of Life and Save the World*.[15]

Centuries! That may be true, but it's not what they were saying in 1999.

The Celera Genomics board fired Craig Venter in January 2002. Things hadn't gone according to plan. Celera's stock peaked early in 2000, at $276 a share. Then it declined sharply, followed by a sharp decline. By the end of September 2005, it was at $11.

Venter got out with $100 million, give or take, and he started a non-profit, the J. Craig Venter Science Foundation. This has freed him to do whatever he wants in the name of science. By the time James Shreeve caught up with him on his yacht in French Polynesia, Venter was "no longer an 'almost billionaire' but was still rich enough to have spent the day shopping on the island for a villa in the $5 million range." He was circling the globe, "having a hell of a good time and getting a very good tan."[16]

Always perceptive about the coming trend, Venter is now saving biodiversity for mankind. He fishes up bits and pieces of exotic flora and preserves their DNA. Someone will run it all through a computer. Harvard's Edward O. Wilson serves on his science advisory board and says flattering things. "This is a guy who thinks big and acts accordingly," he says. "We're talking about an unknown world of enormous importance. Venter is one of the first to get serious about exploring that world in its totality."[17]

Venter himself makes global pronouncements that are likely to be ignored: "We will be able to extrapolate about all life from this survey."

As to those French kids with SCID, the treatment didn't work out because the hit-or-miss viral delivery system can land the new gene anywhere at random. Later analysis showed (in one account) that the incoming gene disrupted another gene, which, when

A Book You're Not Supposed to Read

It Ain't Necessarily So: The Dream of the Human Genome and Other Illusions by Richard Lewontin; New York: New York Review of Books, 2001.

disturbed, can in turn trigger cancer. These potentially cancer-causing genes are called "oncogenes."

It is entrenched dogma within NIH and among most specialists in the field that mutated genes indeed do cause cancer. But there are reasons for thinking that that may not be true. If it is not, we are looking at what may turn out to be the greatest medical error of the twentieth century. It is the subject of the next chapter.

Chapter 11

THE GREAT CANCER ERROR

A book about the gene as cultural icon says that genes "appear to explain obesity, criminality, shyness, directional ability, intelligence, political leanings, and preferred styles of dressing." There are selfish genes, pleasure seeking genes, violence genes, celebrity genes, gay genes, couch potato genes, depression genes, genes for genius, genes for saving, and even genes for sinning.[1]

Scientists tend to believe that almost any trait can be attributed to a gene. The gene obsession, showing up in science journals and on the front page of the *New York Times*, culminated in the Human Genome Project. Now we hear of a Cancer Genome Project. It is supposed to reveal cancer-causing mutations by determining the DNA sequence in thousands of cells from tumor samples. The order of the four letters of the genetic code in these cells, compared with sequences in healthy tissue, would reveal the differences that are known as mutations. There are antecedent, often chemical causes, called carcinogens that are presumed to cause the mutations. But it is the mutations that transform normal cells into cancer cells, according to the mutation theory.[2]

Two things can be confidently predicted if this project goes forward. First, it will cost a great deal of money. One current estimate of $1.35 billion assumes that the cost of sequencing will fall by a factor of ten during

Guess what?

- President Nixon declared a War on Cancer in 1971. But there has been virtually no improvement in cancer mortality.

- For thirty years, the National Cancer Institute has pursued only one theory of cancer's origins.

- Normal human cells have forty-six chromosomes, but cancer cells may have as many as eighty. The theory that aneuploidy is the real cause of cancer was proposed as early as 1914.

the lifetime of the project. Second, the project will not improve our understanding of cancer.

Cancer has remained impervious to every "breakthrough" and treatment hype. Journalists looking for exclusives often view new treatment claims uncritically. Editors search for front-page material. Biotech companies want to create the "buzz" that will encourage investors. And at the National Cancer Institute, in Bethesda, Maryland, any sign of progress is welcomed. Congressional appropriations must be increased! That is the great imperative of government science.

All these things work in the same direction. They disguise how little progress has been made in the War on Cancer, declared by President Nixon in 1971.

"The conquest of cancer is a national crusade to be accomplished by 1976," a congressional resolution declared in 1970. Funds would be made available both for a "massive program of cancer research" and for the "buildings and equipment with which to conduct the research." In the back of everyone's mind was the recent moon landing. Policymakers and scientists tended to assume, then and now, that willpower, optimism, and taxpayers' money was all it took to achieve a scientific goal.

But that's not the way it worked out.

In 1971, when the National Cancer Act became law, 330,000 Americans died of cancer; 650,000 cases were newly diagnosed. The National Cancer Institute's budget was increased—scheduled to reach $800 million a year by 1976.

In 2004, about 560,000 Americans died of cancer, with perhaps 1.25 million newly diagnosed cases. The budget of the National Cancer Institute has reached $5 billion. Other federal agencies contribute another $2 billion, and about $1 billion comes from major charities. The pharmaceutical companies spend $6 billion on cancer R&D.

Fortune published a rare, critical cover story on cancer in March 2004. ("Why We're Losing the War on Cancer.") This was something unusual in journalism, but its author, Clifton Leaf, energized by his own cancer diagnosis years earlier, wasn't buying the usual hype. He is also an executive editor of the magazine. He wrote:

> Even as research and treatment efforts have intensified over the past three decades and funding has soared dramatically, the annual death toll has risen 73 percent—over one and a half times as fast as the growth of the U.S. population. Within the next decade, cancer is likely to replace heart disease as the leading cause of U.S. deaths."[3]

To be sure, the American population is aging, and cancer becomes more prevalent with age. But even adjusting for that, Leaf wrote, "the percentage of Americans dying from cancer is about the same as in 1970 . . . and in *1950*. In contrast, age-adjusted death rates for heart disease and stroke have declined by 59 percent and 69 percent, respectively."

A decline in lung cancer in recent years can be attributed to the decline in smoking, and improvements in five-year cancer survival rates are a function of better surveillance and earlier diagnosis, when surgery has a better chance of success. Survival gains for the more common forms of cancer are measured in additional months of life, not years. As for the billions pouring into cancer research, Leaf wrote, few of these gains result from "exciting new compounds discovered by the NCI labs or the big cancer research centers—where nearly all the public's money goes."

Has something gone wrong? Has the cancer establishment pursued the wrong theory? Since the early 1970s, leading researchers have doggedly pursued one fixed idea about cancer, to the virtual exclusion of all others. The overwhelmingly dominant theory is that cancer is caused by gene mutations. It may prove to have been one of the greatest medical errors of the twentieth century.

Here is the background. In the 1920s researchers bombarded fruit flies with X-rays, and mutant flies resulted. Humans exposed to large X-ray doses a hundred years ago proved to be at high risk for skin cancer and leukemia. So it was convincingly shown that X-rays indeed can produce both mutations and cancers.

Working at the NIH in the 1960s, biochemist Bruce Ames used fast-growing bacteria to detect the mutagenic properties of various substances. Some carcinogens proved to be mutagenic, and here the gene-mutation theory of cancer made its great advance. Robert A. Weinberg, who directs a cancer research lab at MIT, says that by the 1970s he and others had come to believe that "Ames was preaching a great and simple lesson" about carcinogens: Carcinogens are mutagens. They act "by damaging DNA, thereby creating mutations in the genes of target cells."[4]

Some carcinogens indeed are mutagens—radiation, for example—but some of the best known are not. Neither asbestos nor tar, found in cigarettes, is mutagenic. Both are well known as carcinogens, but they don't affect the DNA—the genes.

Viruses were also thought to play an important role. Wendell Stanley, who won the Nobel Prize for chemistry in 1962, and was the first scientist to win the prize in the virus field, said that viruses "hold the key to the secret of life, to the solution of the cancer problem." That was a mistake, soon immortalized in the government's Special Virus-Cancer Program. For viruses to cause cancer, however, cancer would have to be contagious, which it is not.

We Know the Sermon by Heart . . .

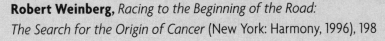

"We knew Bruce Ames's sermon by heart because it was so simple: mutagens = carcinogens. His equation had become the credo of our religion."

Robert Weinberg, *Racing to the Beginning of the Road: The Search for the Origin of Cancer* (New York: Harmony, 1996), 198

As early as 1910, a particular retrovirus *was* shown to cause tumors in chickens. Peyton Rous, who made this discovery, was rewarded with the Nobel Prize in 1966. The virus was named after him. By then, he was eighty-five years old and disillusioned about viruses, because their tumor-causing effect could not be replicated in humans. In his Nobel speech he said that the attempts to demonstrate the carcinogenic effects of viruses more broadly "have consistently drawn blanks, save in a few highly dubious instances."[5]

Nonetheless, officials at the NIH were convinced that viruses played a key role in cancer. In the end, no fewer than five Nobel prizes have been won by researchers who studied tumor viruses.

They include David Baltimore, who today is the president of Caltech; Harold Varmus, who became the director of the NIH under President Clinton and today is president of the Memorial Sloan-Kettering Cancer Center in New York; and J. Michael Bishop, today the Chancellor of the University of California in San Francisco, one of the leading medical research institutions in the country.

In short, three scientists who are today among the most influential in the country won the Nobel Prize for their study of a chicken virus— which its discoverer did not think was a good model for cancer. Few seem to appreciate the irony.

In 1970, a young scientist at Berkeley, Peter H. Duesberg, toiling away in the platoons of researchers hunting for viral cancer, discovered a gene in the Rous virus. He found that it *was* capable of transforming cells into tumor cells, and could do so in a matter of days. He had identified what became known as the first "cancer gene," called *src*. (Pronounced "sarc," an abbreviation of sarcoma).

At the same time, two researchers at the NIH, Robert Heubner and George Todaro, put forward the "oncogene hypothesis." Oncology is the study of cancer, and the oncogene hypothesis attributed all cancer to the "activation" of certain genes that are intrinsic to cells (not transported there by viruses). These hypothetical cancer genes, known as *proto-oncogenes*, are "activated" by being mutated.

Duesberg was rewarded with membership in the National Academy of Sciences. But further study gradually convinced him that the viral *src* gene was a rare and unstable biological accident, not relevant to human cancer. In that he agreed with Peyton Rous.

Thus began his career as perhaps the best known dissident scientist in the country. Later, in 1986, when the editor of a leading cancer journal asked him to take an independent look at AIDS, he concluded that it was not infectious and was not caused by HIV.[6] Not long after that all his NIH funding was cut off. and it has never been restored. Later still, Duesberg would make an important cancer discovery.

But across the San Francisco Bay, Bishop and Varmus took the hint from the NIH and looked for signs of the *src* gene in normal cellular tissue. Describing their work in his book *How to Win the Nobel Prize*, Michael Bishop wrote: "It required the better part of four years before we reached the conclusion that vertebrate cells do indeed carry a version of *src*."[7]

A *version of*, notice. It was not the same thing—an important distinction in a field that was on the verge of claiming that a *single point mutation* can transform a normal gene into a cancer gene.

So Bishop and Varmus found *something like* the src gene in vertebrates, including human cells, without any trace of the virus. Their discovery was reported in *Nature* in 1976.

The theory was that this gene, when mutated, caused cancer. Thirteen years later, in 1989, they were rewarded with the Nobel Prize. The *New York Times* reported:

> 2 Doctors Share Nobel Prize for Work With Cancer Genes;
> Experiments with Chicken Cells Lead to a New Era in Research

"Their research showed that normal genes that apparently control cell growth can undergo alterations that lead to the uncontrolled growth that is cancer," Gina Kolata reported. She added this detail: their "prize-winning research has led to the discovery of more than forty different genes that can cause cancer." Bishop, Varmus, and others "have found that the genes can cause cancer in people as well as animals."[8]

A week later, *Science* reported:

Cancer Gene Research Wins Medicine Nobel

"The Bishop-Varmus group has had a major impact on efforts to understand the genetic basis of cancer," Jean Marx wrote. "Since their 1976 discovery, researchers have identified nearly fifty cellular genes with the potential of becoming oncogenes."[9] By the time Michael Bishop wrote *How to Win the Nobel Prize*, the repertoire of oncogenes had been expanded "to one hundred or more."

Then, in 2004 the *Washington Post* reported on its front page:

Genetic Test Is Predictor of Breast Cancer Relapse

The test "marks one of the first tangible benefits of the massive effort to harness genetics to fight cancer," Rob Stein wrote.[10] Almost thirty years after Bishop and Varmus's first investigation, little had been achieved. Two well-publicized genes, *brca-1* and *brca-2*, supposedly predispose women to breast cancer, but in over 90 percent of all breast cancers these genes have shown no defect. And most women who carry this gene in mutated form do not get breast cancer.

The *Post*'s Stein added that researchers had "sifted through 250 genes that had been identified as playing a role in breast cancer." The Sanger Institute in Britain, which participated in the Human Genome Project, claimed recently that "currently more than 1 percent of all human genes are cancer genes." The latest gene total for humans is 25,000, and that is where the 250 "cancer genes" came from.

Notice the multiplication of genes "playing a role" in cancer. Once it was just the *src* gene. Now there are 250. This "multiplication of entities" is the hallmark of a theory that is not working. It is what philosophers call a "deteriorating paradigm." The theory gets more and more complex to account for its lack of success. The number of oncogenes keeps going up, even as the number of human genes actually counted goes down. Six years ago some thought we had 150,000 genes. Now it's one-sixth that number. How long before they find that *all* the genes "play a role" in cancer?

Furthermore, the oncogene theory at the outset posited that a *single gene mutation* was sufficient to turn a normal cell into a cancer cell. But that always was unlikely. Mutations occur at a predictable rate in the body. As the cells of the body number in the tens (or maybe hundreds) of trillions, we would all have cancer if a single hit was sufficient to transform a cell.

In time, therefore, the theory was adjusted. "Two hits" and then "multiple hits" were posited. Maybe six or seven genes would all have to mutate in the same cell during its lifetime. Then, bingo, your unlucky number had come up. That cell became a cancer cell. But that was all little more than guesswork based on gene mutation theory.

There are several problems with the theory, but the most basic is this. Researchers were never able to show that the original *src* gene, or any other gene, whether or not mutated, could transform a cell into a cancer cell. Nor could it start a tumor in any animal. Furthermore, they have not been able to show that any *combination* of genes taken from a cancer cell can transform normal cells, when tested in vitro, or in the lab. The theory has never been confirmed by any functional test.

In short, they have never been able to show that the allegedly guilty party is capable of committing the crime. They can indeed transport these mutated genes into test cells. And these genes are integrated into the cell's DNA. But the recipient cells *do not turn into cancer cells*, and if injected into experimental animals, they *do not cause tumors*.

That's when the experts said, well, there must be four or five genes all acting at once. But they have never been able to say which ones, nor show that in any combination they do the foul deed. There is even a genetically engineered mouse strain called OncoMouse, and many more have been generated since then. These genetically engineered mice have mutated oncogenes in every cell of their small bodies. You would have thought they would die of cancer immediately. But they leave the womb, gobble up food, and live long enough to reproduce and pass on their supposedly deadly genes to the next generation.

So how are these hundred or more oncogenes identified? It has reached the point where researchers hunt through the DNA of clinical cancer cells until they find a mutated gene. At that point they feel entitled to claim that they have discovered a gene that "played a role" in the transformation of that cell. One more "oncogene" may then be tallied. Association is raised to the level of causation. Bystanders become culprits.

The desire to start over with a "cancer genome project" itself tells you that the gene mutation theory hasn't gone anywhere. Some of the researchers involved may even know it. Dr. Harold Varmus told the *New York Times* in March 2005 that the new project could "completely change how we approach cancer."[11]

Completely change? That's not likely, but maybe we do need a complete change. What about his thirty-year-old Nobel work? Was that all a waste? It may have been worse than that, because when a misleading theory is rewarded with the top prize in science, abandoning the theory becomes extremely difficult. Rival theories are likely to be discarded. The backtracking required is an embarrassment to all.

There was one crucial discovery still to be made—or rather, a rediscovery. Again it was made by Peter Duesberg, and as recently as 1997. Over a quarter of a century had passed since his elucidation of the *src* gene.

In 1982, Robert Weinberg of the Whitehead Institute at MIT claimed that a mutation in a single gene indeed *had* transformed a cell in vitro. It was called the *ras* gene. But it turned out that the cell-line into which it was transferred—it had been provided by the NIH, and is known as the

3T3 cell—was "immortalized." Cells in the body normally divide only a certain number of times. They obey what is called the Hayflick limit—named after Leonard Hayflick of UCSF, who discovered that most cells divide about fifty times before expiring. It is thought to be a cause (or a concomitant) of aging.

Immortalization "rang an alarm among the scientists reexamining the gene transfer experiments," as Weinberg wrote in his 1998 book *One Renegade Cell*. And in a significant aside, he wrote: "Curiously, virtually all types of cancer cells also seem to be immortalized." That, incidentally, is why they are so dangerous: once they get going, they never quit.[12]

What had happened was that Weinberg's research had been conducted with cells that were already cancerous. A great deal of in vitro research over the years was done with these NIH 3T3 cells, creating the impression that when they became overtly cancerous it was the transferred genes and not the initial condition of the cell that had done the trick.

What was the salient structural feature of these immortalized cells? They did not have the right number of chromosomes. Normal mouse cells have forty chromosomes, humans forty-six—twenty-three each from mother and father. Such cells are called "diploid," because they have two of each chromosome. Genes are segments of DNA strung along these chromosomes. The largest chromosomes incorporate several thousand genes each. Sometimes, babies are born with one extra copy of the smallest chromosome, and because this defect is in the germ line, it is found in every cell of the body. Such babies have Down's Syndrome. Having an extra chromosome is serious business, and these babies only survive at all because the "trisomy"—the tripled chromosome—is the smallest chromosome.

Here is the discovery that Peter Duesberg made, when he began to look back through the research buried in the vast cancer literature. In every case where the information was included, cancer cells turned out to have an incorrect complement of chromosomes—usually too many. Their "ploidy" is not good, so they are said to be (in the Greek derivation) *aneuploid*. He theorized that aneuploidy was the real cause of cancer.

In cancer, it is important to stress, this defect arises *not* in the germ line, but in the grown body. Cells divide in the course of life, sometimes frequently (in the gut, in the skin) sometimes infrequently (bone, muscle), and they divide by a process called mitosis. The chromosomes double up, and then segregate, and after mitosis a full complement of chromosomes ends up in both "daughter" cells.

But sometimes there is an error in the separation. The chromosomes do not divide properly, but end up unequally in the daughter cells. An extra chromosome may be hauled off into one of the new cells. One cell will then come up short; the other will have a surplus. Such under-endowed or over-burdened cells will usually die, and then there is no

It All Started with a Chicken Virus . . .

"When I came to the United States, in 1964, the Special Virus Cancer Program had just been launched. The dominant theory was that viruses cause cancer. I worked on that at Berkeley under Wendell Stanley—the first virologist to win the Nobel Prize.

"Decades earlier, Peyton Rous had discovered that a particular retrovirus causes a tumor in chickens. But it proved to be an unstable, biological accident not relevant to human cancer. If viruses really caused cancer, cancer would be contagious, and it is not.

"Then, in the 1970s, something resembling the cancer gene of the Rous virus was found in normal human DNA. At that point, it was claimed that a human 'oncogene' had been discovered. In mutated form, it would turn the cell into a cancer cell—in theory.

"Today, over one hundred oncogenes and dozens of 'tumor suppressor' genes have been discovered. They are said to be 'associated with' cancer; but not one, nor any combination of them, whether mutated or not, has been shown to transform a normal cell into a cancer cell."

Peter Duesberg, author interview, 2005

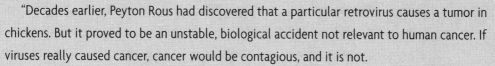

problem. But sometimes the error persists, repeats, magnifies, and increases, more likely in the cell with an extra chromosome. The cell just keeps on dividing, ignoring the Hayflick limit. Its control mechanisms are overridden. And this is dangerous.

A tumor forms in that part of the body, and that is cancer. Some human cancer cells may have as many as eighty chromosomes instead of forty-six. They may actually have double the right number of genes. It's not just a matter of a cell with two or three genes that malfunction. It is far worse than that. And the more aneuploid they become, the more likely they are to metastasize.

When researchers began to look at cancer cells under the microscope about a hundred years ago, the first thing they noticed was the abundance of chromosomes. One who made this discovery was Theodor Boveri, who wrote a book about it in 1914.[13] Early researchers thought this surely was the genetic cause of cancer. Mutations didn't enter into it. Peyton Rous didn't mention mutations in his 1966 Nobel acceptance speech.

But here was the historical problem. The cells of tumors initiated by the Rous sarcoma virus *did* have the correct set of chromosomes. They were diploid. Researchers who had been studying them for many years therefore became convinced that the "ploidy," or chromosome count of cancer cells, was irrelevant. But when Duesberg reexamined these virally induced tumors later, he concluded that they were not true cancers at all. They were reversible. They were more like warts. When the chicken's immune system kicked in, the tumors disappeared.

Gradually the early research by Boveri and others was forgotten. Then it was buried in the endless blizzard of new research. In the last generation, textbooks on the cell and even textbooks on cancer have failed to mention aneuploidy or its bizarre chromosomal combinations. Robert Weinberg wrote two books on cancer without mentioning aneuploidy. And Theodor Boveri's book cannot even be found in the National Library of Medicine, on the NIH campus in Bethesda. Overlooking what was plainly visible in the microscope, researchers worked for years with defective, immortalized cell lines, assuming their extra chromosomes were irrelevant.

Duesberg uses a metaphor to contrast the two theories. Cells are often compared to factories, so think of an automobile plant instead of a cell. A cancer cell is the equivalent of a monster car with (say) five wheels, two engines, and no brakes. Start it running and you can't stop the thing. It's a hazard to the community. The CEO of the auto company wants to know what is wrong, so he sends investigators into the factory. There they find that instead of the anticipated forty-six assembly lines, there are as many as eighty. And at the end of the process this weird machine gets bolted together and then plows its way out of the factory door. So that's the explanation, right? All those extra assembly lines!

But today's gene mutation theorist is someone who says: "That's not it. The extra assembly lines are irrelevant. What happened is that three or four or five of the thousands of workers along the assembly lines are not working right!"

Any CEO would fire the assistant who thought a few errant workers and not the bizarre factory layout had caused the mayhem. But in the realm of cancer research, those who do say that are rewarded with fat grants, top posts, and Nobel Prizes. That's a measure of what has happened to cancer research. Today's leading cancer

Cancer Cell as Deranged Automobile Factory

"All cancer cells have the wrong number of chromosomes, or the wrong complement of them. If the cell is likened to an automobile factory, a cancer cell may have sixty or even eighty assembly lines (chromosomes) instead of the normal forty-six. In the end, a weird vehicle with five wheels, two engines, and no brakes gets bolted together. It's a hazard to the community and can't be stopped. What is the cause? The wrong number of assembly lines! But NIH-approved cancer researchers say "no, that's not it." They are convinced that three or four of the tens of thousands of workers strung along the assembly lines are not working right. Bad workers, of course, are the equivalent of mutated genes."

Peter Duesberg,
author interview, 2005

researchers still do not habitually look at the chromosome combinations found in the cells they work with.

Harold Varmus at Sloan-Kettering said in an e-mail reply to my query: "Aneuploidy, and other manifestations of chromosomal instability, are major manifestations of many cancers and many labs have been working on them." But: "Any role they play will not diminish the crucial roles of mutant proto-oncogenes and tumor suppressor genes."

Why not? Maybe aneuploidy can do the job on its own?

By the time that Duesberg noticed the aberrant chromosomes, omnipresent in the cancers he was studying, he was already in the dog-house at NIH. First he had said that the *src* gene was not relevant to human cancer, then that HIV is harmless. All his NIH grants had been cut off. He was sixty years old. He was assigned to teach an undergraduate lab course. But he still had tenure at Berkeley, and as a member of the National Academy he could still publish in respected journals. (But his contribution to the Proceedings of the National Academy of Sciences, unlike those of other Academy members, have been minutely scrutinized by peer reviewers.)

For the last seven years Duesberg has been drawing attention to the cancer issue. The NIH is pursuing the wrong theory of cancer origins, he says. All his subsequent grant applications have been turned down.

A researcher at the University of Washington who became controversial at NIH in an unrelated field warned Duesberg: "In the present system of NIH grants, there is no way to succeed." No matter how much they proclaim that they want applicants to "think outside the box" and make "high-risk" grant proposals, they really don't mean it. "The reviewers are the same and their self-interest is the same." In the cancer field, grant proposals are both reviewed by and won by proponents of the gene mutation theory.

W. Wayt Gibbs published a good article about Duesberg's cancer findings in *Scientific American*.[14] Duesberg's work is not being ignored, and it will not be, because we know that what was often said of AIDS (but was not true) really is true of cancer: We are all at risk.

The following response is often heard from the upper echelons of cancer research: "We have *always known* that aneuploidy is important in cancer." (Yes, but it was forgotten and then buried beneath the mountains of new research.) A quiet search is under way for a congenial "political" compromise. (Some researchers are beginning to say: "Maybe both gene mutation *and* aneuploidy 'play a role' in cancer.")

A leading cancer researcher, Bert Vogelstein of Johns Hopkins, accepted a few years ago that "at least 90 percent of human cancers are aneuploid." More recently, his lab reported that aneuploidy "is consistently shown in virtually all cancers." More and more researchers accept that, yes, cancer cells are aneuploid. But some insist that this is a consequence, not a cause of cancer. But then Christoph Lengauer, a researcher in the Vogelstein lab, said that "with our experiments, we found that aneuploidy is a very early event in tumorigenesis." Believing it was a consequence of cancer was "a major mistake the scientific community made."[15]

Nonetheless, Bert Vogelstein has been closely identified with the earlier discovery of oncogenes and "tumor suppressor" genes, and he won't abandon the mutation theory easily.

A Book You're Not Supposed to Read

"Retroviruses as carcinogens and pathogens: expectations and reality" by P. H. Duesberg, *Cancer Research*, v. 47, 1987. Reprinted in P. H. Duesberg, *Infectious AIDS: Have We Been Misled?* Berkeley: North Atlantic Books, 1995.

Recently, he has been looking for mutations that cause aneuploidy. That is undoubtedly the direction in which a compromise will be sought.

At the end of May 2005, Duesberg was invited to speak at NIH by the Cancer Genetics branch of the National Cancer Institute. His topic: "Aneuploidy and Cancer: From Correlation to Causation." About a hundred people showed up at Building 10. The Genetics branch of the National Cancer Institute is interested in aneuploidy and is aware of the political sensitivities. Thomas Ried, the head of the Genetics division, is well familiar with Duesberg's arguments.

Duesberg challenged the audience to prove him wrong. He is looking for diploid cancer, he said: a true cancer with the correct set of chromosomes. He is not much interested in compromise solutions, and this is the attitude that made him both politically unpopular and a true scientist. Prove me wrong, he says. Give me a reference to diploid cancer in the scientific literature. No unambiguous case has yet been reported, he claims.

The medical implications are not easily fathomed. The aneuploid basis of cancer is not encouraging, in the sense that no foreign or outside elements are involved in the cancer cell. Therefore, it will not be caught by vaccines or by the immune system. On the other hand, it may be that the abundant DNA of the most aberrant cells will make them traceable. Extra chromosomes are much easier to see than DNA misprints, so it should make diagnosis easier. And, in fact, aneuploidy is already being used as a diagnostic guide in Germany and Scandinavia.

More generally, we may say that the experts will never find a solution to cancer if they continue working with the wrong theory, as it seems they have been with gene mutations. With the right theory, a dramatic breakthrough may really be achieved.

As to the interaction of science and politics, Peter Duesberg may in the end have shown that in order to "think outside the box," it is better not to be funded within the box. If so, we will all have learned a very expensive but indispensable lesson.

Chapter 12

THE ABIDING MYTHS
Flat Earth and Warfare between Science and Religion

The idea that educated people believed in a flat earth for centuries is legendary in the annals of science. The legend lives on, though in a new way. A literal believer in a flat earth was someone who espoused an indubitably false theory. If you sailed across the ocean far enough, you might fall off. No one believes that today, yet people are still described as "flat earthers." It has become a term of derision, an ad hominem accusation directed against someone who *casts doubt* on a particular scientific theory—often a new theory—that may indeed merit suspicion.

Traditionally, doubting new theories, especially those that have been promoted for political reasons, has been an essential part of the scientific enterprise. Whenever the "flat-earther" label is brandished today, you can be sure that what is at stake is some politicized claim masquerading as established scientific truth.

In fact, the epithet is frequently aimed at the growing number of people who question evolution. This is appropriate because it is precisely in that context that the myth of a flat earth was first publicized. It was an early warning shot in the evolutionist wars.

The myth is often used to imply that knowledge, once secure in the pre-Christian era, was lost in the "Dark Ages." According to this misreading of history, learning only got back on its feet at the time of the Enlightenment,

Guess what?

‧ The claim that medieval scientists and theologians believed the earth is flat was concocted in the nineteenth century.

‧ Galileo, one of the first casualties in the alleged war between science and religion, could have avoided trouble with the Catholic Church if he had stuck to science and not ventured into theology.

‧ Christianity elevated the faculty of human reason and fostered a spirit of inquiry. Without it, there would never have been a scientific revolution.

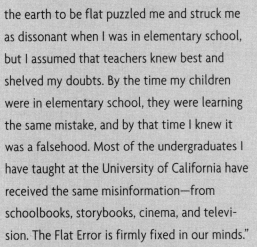

School Days

"The almost universal supposition that educated medieval people believed the earth to be flat puzzled me and struck me as dissonant when I was in elementary school, but I assumed that teachers knew best and shelved my doubts. By the time my children were in elementary school, they were learning the same mistake, and by that time I knew it was a falsehood. Most of the undergraduates I have taught at the University of California have received the same misinformation—from schoolbooks, storybooks, cinema, and television. The Flat Error is firmly fixed in our minds."

Jeffrey Burton Russell,
Inventing the Flat Earth;
Westport, CT: Praeger,
1991, xiii.

when "superstition"—itself usually a code word for religion—began to decline. Science, opposed and hobbled for centuries by its ancient enemy, religion, at last regained the upper hand. The myth of the flat earth is best thought of as one aspect of the broader theme that religion has long been making war on science. And that, too, is a myth.

Perhaps the most eminent modern proponent of the idea that scientific understanding was eclipsed for a millennium was Daniel Boorstin, a former librarian of Congress. In his 1983 bestselling book *The Discoverers*, he wrote, "A Europe-wide phenomenon of scholarly amnesia... afflicted the continent from AD 300 to at least 1300. During those centuries Christian faith and dogma suppressed the useful image of the world that had been so slowly, so painfully, and so scrupulously drawn by ancient geographers." He calls this period the "Great Interruption."[1]

There is just one problem. It's a myth. But it is still sometimes taught in high schools and colleges. "How could a better story for the army of science ever be concocted?" said Stephen Jay Gould, explaining how the flat-earth myth was exploited. Supposedly, our ancestors "lived in anxiety, restricted by official irrationality, afraid that any challenge could only lead to a fall off the edge of the earth into eternal damnation." But that tale is "entirely false," Gould added, "because few medieval scholars ever doubted the earth's sphericity."[2]

The man who most effectively exposed the myth he calls the "Flat Error" is Jeffrey Burton Russell, an emeritus professor of history at the University of California. His book *Inventing the Flat Earth*, published in 1991, went a long way toward correcting the error. Russell summarizes the early understanding as follows:

> In the first fifteen centuries of the Christian era [only] five writers seem to have denied the globe, and a few others were ambiguous and uninterested in the question. But nearly unanimous scholarly opinion pronounced the earth spherical, and by the fifteenth century all doubt had disappeared. There was no 'Great Interruption' in this era."[3]

Earlier, C. S. Lewis, an expert on Renaissance literature in addition to being a popularizer of Christian theology, wrote that "physically considered, the earth is a globe; all the authors of the high Middle Ages are agreed on this... The implications of a spherical earth were fully grasped."[4]

During the early and medieval Christian eras, the leading lights of the age, Saint Augustine, the Venerable Bede, Saint Thomas Aquinas, Roger Bacon, and Dante all affirmed a spherical earth. So did "the greatest scientists of later medieval times," as Gould identifies them, Jean Buridan (1300–1358) and Nicholas Oresme (1320–1382).

How did so gross an error arise and become a part of our mental furniture? When he began to look into the question, Russell was sure he would find the same idea echoing back through the centuries. But he was surprised to find that even the philosophers of the French Enlightenment, with their strong anti-Christian bias, had hardly written a word on the subject. Gould accepts that "none of the great eighteenth-century anticlerical rationalists—not Condillac, Condorcet, Diderot, Gibbon, Hume, or our own Benjamin Franklin—accused the scholastics of believing in a flat earth."

The earliest promoter of the myth turns out to be none other than Washington Irving, the creator of Rip Van Winkle. In 1828, Irving wrote the largely fictitious *History of the Life and Voyages of Christopher Columbus*, setting the stage by fantasizing a confrontation at the university in Salamanca in 1491. Columbus appeared, he wrote, as "a simple mariner, standing forth in the midst of an imposing array of professors, friars, and dignitaries of the church, maintaining his theory [of a spherical earth] with natural eloquence and, as it were, pleading the cause of the new world."[5]

Historian Samuel Eliot Morison later described Irving's account as "pure moonshine." It was "misleading and mischievous nonsense."

Columbus does play an important role in the flat-earth myth. He not only discovered America, but (supposedly) amazed his contemporaries by proving that the earth is round. "It is an illusion by no means confined to the uneducated," Russell writes.

By the nineteenth century, Christopher Columbus had been transformed into a bold rationalist who overcame ignorant churchmen and superstitious sailors. It is a portrait without any basis. He was less a rationalist, Russell writes, than "a combination of religious enthusiast and commercial entrepreneur."

It is said that before setting forth on his voyage, Columbus had to face down bigoted clerics who warned that his boat would fall off the earth. Objections to the voyage were indeed raised at Salamanca, but none on the grounds that the Earth was flat. In fact, one objection was that Columbus may have greatly underestimated the distance across the ocean (to Japan)—as he had, by a considerable margin. None of the early historical sources on his voyage say anything about a dispute regarding the earth's roundness.

The true "Flat Earthers"

The two most important salesmen of the flat-earth myth were nineteenth-century Americans. John William Draper's *History of the Conflict between Religion and Science* was published in 1874, and Andrew Dickson White's tract *The Warfare of Science* was expanded two decades later into

A History of the Warfare of Science with Theology in Christendom.

A disaffected Methodist, Draper (1811–1882) was born in England and came to the U.S. when he was twenty-one. He became a professor of chemistry and head of the medical school at New York University. White (1832–1918), a New York state senator, founded Cornell University in 1865 as the first explicitly secular university in America. He became president of Cornell by the age of thirty-three.

Draper used the flat-earth myth to illustrate his thesis that the Catholic Church was antagonistic to learning. "The writings of Mohammedan astronomers and philosophers had given currency to that doctrine [of a spherical earth] throughout Western Europe, but as might be expected, it was received with disfavor by theologians," he wrote. "Traditions and policy forbade [the papal government] to admit any other than the flat figure of the earth, as revealed in the Scriptures."

White wrote in *History of the Warfare*:

> Many a bold navigator, who was quite ready to brave pirates and tempests, trembled at the thought of tumbling with his ships into one of the openings into hell which a widespread belief placed in the Atlantic at some unknown distance from Europe. This terror among sailors was one of the main obstacles in the great voyage of Columbus.[6]

The strong American involvement is striking. Perhaps what we see here is an unanticipated consequence of the separation of church and

The Christian Tradition

Edward Grant, a leading professor of the history and philosophy of science, said of the supposed repression of reason in the Middle Ages: "If revolutionary rational thoughts were expressed in the Age of Reason they were only made possible because of the long medieval tradition that established the use of reason as one of the most important of human activities."

God and Reason in the Middle Ages; Cambridge: Cambridge University Press, 2001.

state. For where there is no "entanglement" with the state, religion can be opposed without fear of state reprisals, and where the press is free, as it was (and still is) to a far greater extent in America than in Europe, a more vigorous antagonism could be freely expressed.

There were considerable differences between the two authors, however. Concentrating almost exclusively on Roman Catholicism, Draper took the opportunity to excoriate the Church for its supposed opposition to science. He was particularly agitated by Pius IX, who issued the Syllabus of Errors and the declaration of papal infallibility in 1870.

After attaining political power in the fourth century, Draper wrote, the Church had displayed "a bitter, a mortal animosity" toward science, and its persecution of scientists had left its hands "steeped in blood." Islamic scholars, on the other hand, had laid the foundations of several sciences and the Protestants had cultivated a "cordial union" with science marred only by occasional misunderstandings. Draper concluded:

> Religion must relinquish that imperious, that domineering position which she has so long maintained against Science. There must be absolute freedom for thought. The ecclesiastic must learn to keep himself within the domain he has chosen, and cease to tyrannize over the philosopher, who, conscious of his own strengths and the purity of his motives, will bear such interference no longer.[7]

White was more moderate than Draper, and he became more so over the decades that he addressed the supposed conflicts of science and religion. In a lecture at New York's Cooper Institute in 1869, he sounded the themes he would later develop. Whenever religious dogma obstructs the "liberty of science," said White, both science and religion suffer; they experience "the direst evils." On the other hand, "all untrammeled scientific investigation, no matter how dangerous to religion some of its stages may have seemed," has invariably "resulted in the highest good of religion and science."[8]

A similar claim is made by bioengineers today. White had no notion that the scientific enterprise he saw around him operated within the self-imposed constraints of a largely Christian culture, or that the beneficial effects of such constraints can easily be taken for granted. When removed in the twentieth century, the new value-free environment could be exploited by the likes of Dr. Josef Mengele.

White gradually narrowed the focus of his attack. He began by criticizing religion, then "ecclesiasticism," and finally "dogmatic theology." The latter "smothered" truth, he decided. A tendency to dogmatism "which has shown itself in all ages" is "the deadly foe not only of scientific inquiry but of the higher religious spirit itself."[9]

Here, White sounded a very modern note. Religion per se was not necessarily the problem because it could yield and accommodate, as the Unitarians and then the Anglicans had shown in White's (and in Darwin's) lifetime. At best, religion consisted merely of recognizing a "power in the universe" and living by the Golden Rule. The real problem was "dogmatism," which was unbending.

Both Draper and White developed their ideas about the warfare between science and theology in the context of the emerging debate about evolution. Darwin's *Origin of Species* had been published in 1859, and both authors sought to promote the cause of evolution. No issue before or since has so deeply challenged traditional views concerning the meaning of life.

Draper himself had participated in what may well have been the first skirmish in the war between Darwin and divinity. In 1860, a famous debate between Bishop Wilberforce and Darwin's aggressive young supporter, Thomas Henry Huxley, was held at the Museum of Zoology in Oxford. In their oft-discussed exchange, Wilberforce is said to have asked

Huxley if he was descended from an ape on his grandmother's or his grandfather's side. Huxley said he preferred to be descended from an ape than a bishop. (No stenographer was present, the acoustics were poor, and what they really said is much disputed.)

The formal paper introducing the debate was delivered by John Draper. His topic was "the intellectual development of Europe considered with reference to the views of Mr. Darwin." His address, expanded into a book, argued that humanity was making slow but steady progress toward a more advanced state under the guidance of science. It expressed the emerging belief in progress, already gathering steam by the 1850s and destined to endure until the First World War.

In her 1957 biography of Darwin, Ruth Moore captured the aftermath of the Oxford debate. It still holds true today: "From that hour on, the quarrel over the elemental issue that the world believed was involved, science versus religion, was to rage unabated." Or as Stephen Jay Gould said, "The Darwinian revolution directly triggered this influential nineteenth-century conceptualization of Western history as a war between . . . science and religion."

Draper's book went through fifty printings in fifty years and became, Gould wrote, "the most successful of nineteenth-century publishing projects in popular science." Jeffrey B. Russell believes it was the first book by an influential figure who "explicitly declared that science and religion were at war." As to its influence, "it succeeded as few books ever do. It fixed in the educated mind the idea that 'science' stood for

Those Bad Middle Ages

A well-known college textbook, *A History of Civilization: Prehistory to 1715*, informed students (in editions published from 1960 to 1976) that although the ancient Greeks knew the Earth was spherical, this knowledge was lost in the Middle Ages. A 1983 textbook for fifth graders reports that Christopher Columbus "felt he would eventually reach the Indies in the East. Many Europeans still believed that the world was flat. Columbus, they thought, would fall off the Earth."

freedom and progress against the superstition and repression of 'religion.' Its viewpoint became conventional wisdom."

But White's *Warfare* proved to be the more influential, the American historians of science David Lindberg and Ronald Numbers argue, partly because "Draper's strident anti-Catholicism soon dated his work and because White's impressive documentation gave the appearance of sound scholarship." Well into the twentieth century, these scholars add, "militaristic language dominated discussions of science and religion, especially during the 1920s.... As late as 1955, Harvard's distinguished historian of science, George Sarton, was still praising White—and suggesting that his thesis be broadened to include non-Christian cultures."[10]

Russell concludes that with extraordinarily few exceptions, "no educated person in the history of Western Civilization from the third century BC onward believed that the earth was flat." But many educated persons since the time of Darwin have wanted to believe otherwise. Their acceptance and promotion of this myth gave them a weapon of ridicule in the war against religion.

The warfare myth

White's book inspired British philosopher Bertrand Russell to write *Religion and Science* in 1935. He believed there had been a prolonged conflict between religion and science over the centuries "waged by traditional religion against scientific knowledge." Religion had the upper hand, then science "after fitful flickering existence among the Greeks and Arabs, suddenly sprang into importance in the sixteenth century."[11]

The Greeks knew what was what and the Arabs had been wise enough to learn from them—that became the approved storyline. But a time of darkness ensued—the Dark Ages. It was the time of Christian domination. But by the sixteenth century, liberation was at hand. Science began to challenge the church's authority. But the clergy fought back, using the

Inquisition and the stake. Nonetheless, reason prevailed. Learning was revived. Only for so long could man be prevented from thinking freely.

Russell devoted his second chapter to the Copernican Revolution—the first "pitched battle between theology and science." Was the earth or the sun at the center of the solar system? It was an old question. The Greeks had asked it early on, and had answered it. But the knowledge was lost

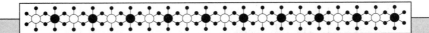

What They Don't Want to Know

Before the twentieth century, few scientists saw any conflict between religion and science. In his "Thoughts on Religion," Professor G. J. Romanes said in the nineteenth century that one thing influenced him in his return to faith. At Cambridge University, nearly all the eminent scientists were avowed Christians. "The curious thing," he says, "is that all the most illustrious names were ranged on the side of orthodoxy. Sir W. Manson, Sir George Stokes, Professors Tait, Adams, Clerk Maxwell, and Bayley...were all avowed Christians."

The pioneering geneticist Gregor Mendel was a monk, Louis Pasteur a devout Catholic. Among those who made great advances in nineteenth-century science "were not a few sons of the Church," wrote James R. Moore in his study of the post-Darwinian controversies: "Galvani in electricity, Fresnel and Fraunhofer in optics, LeVerrier in astronomy." The names of Andre-Marie Ampere (amps) and Alessandro Volta (volts) could be added, both religious men.

David Lindberg, a professor of the history of science at the University of Wisconsin, wrote (with Ronald Numbers): "Despite a developing consensus among scholars that science and Christianity have not been at war, the notion of conflict has refused to die." In its traditional form, John Brooke wrote, the warfare hypothesis "has been largely discredited." Steven Shapin, a professor of the history of science at Harvard, says that though "it was common" in the late Victorian period to write about the "warfare between science and religion," it has been "a very long time since these attitudes have been held by historians of science."

for almost 2,000 years. It was hidden by the church, using its worldly power to keep us in ignorance. Then a free spirit emerged, Copernicus, who rediscovered the heliocentric system.

How did the Copernican theory "threaten" the Church? "The dethronement of our planet from its central position suggests to the imagination a similar dethronement of its inhabitants," Bertrand Russell wrote.[12] It was hard to believe that a remote corner of the cosmos "could have the importance to be expected of the home of Man, if Man had the cosmic significance assigned to him in traditional theology. Mere considerations of scale suggested that perhaps we were not the purpose of the universe."

This claim has been often repeated, with Stephen Jay Gould in particular reveling in the alleged humiliation of man by his removal to a corner of the cosmos.

The argument is false, for a number of reasons. The claim that one part of the universe is more central than another has little meaning. Albert Einstein instructed us that the universe has "no hitching post" or absolute reference point. One thing can be said to move only in relation to another, and no vantage point is privileged.

The history is wrong, too. The critics of Copernicus were not concerned about dethronement. John Wilkins, the leading English Copernican and master of Trinity College at Cambridge when Isaac Newton was a student there, said that the prevalent objection to the Copernican system wasn't man's dethronement but his improper elevation. The center of the cosmos was not a good place to be. "It was the point to which earthly matter fell, the focus of change and impurity, and the physical correlate of humanity's fallen state," wrote John Hedley Brooke, a professor of science and religion at Oxford. "To be placed on a planet was to move upmarket."[13]

Galileo himself was conscious of this, rejoicing that in our newfound orbital existence "there was an escape from the refuse." Among the Copernicans the dominant sentiment was pride rather than humiliation, for they knew that their understanding of the cosmos exceeded that of the ancients.

The more important question is whether the church's war on science has any historical validity. The case of Galileo, who dared to peer into the heavens with his telescope and was hauled before the Inquisition, has been repeated ad nauseam. We hear of little else in the retelling of the so-called war, and for a good reason: there is little else to tell. It was more an isolated skirmish than a pitched battle.

Another side to the story

Nicholas Copernicus, born in 1473, was canon of a Catholic cathedral in Poland. He studied mathematics, physics, and astronomy at the University of Bologna—a Catholic university—where his teacher criticized the earth-centered Ptolemaic system on the grounds of excessive complexity. In about 1514, Copernicus summarized his conclusions in his *Little Commentary*. The planets revolved around the sun, he said, and the combined orbital and rotational motion of the earth could explain the apparent motions in the heavens.

Pope Leo X, informed of the theory, expressed interest, and for a time the heliocentric hypothesis won papal favor. Not so among some of the leaders of the Reformation, however. In Geneva, John Calvin opposed Copernicus, and Martin Luther rejected him with his customary vehemence: "This fool wishes to reverse the entire scheme of astronomy." The Copernican opus, *On the Revolutions of the Celestial Orbs*, was finally published in 1543. One of the first copies, tactfully dedicated to Pope Paul III, reached Copernicus on his deathbed. Throughout his life, he was not even remotely persecuted.

A preface by a Lutheran minister explained that the ideas in the book were only theory. Yet the work as a whole could be regarded as true if its premises "lead to a computation that is in accordance with the astronomical observations." This very modern formulation of scientific truth, from a clergyman, has been strangely criticized as bowing to the Vatican, but it was entirely reasonable and justified in light of history. Some of Copernicus's postulates—for example, his claim that the planets move in circles and that the sun is immobile at the center of the universe— have been discarded. The planets move in ellipses and the sun has its own motion.

Seventy years pass before we hear of Galileo Galilei (1564–1642). He made his first telescope in 1609 and published what he saw in *Sidereal Messenger*. He soon found that his Jesuit friends in Rome had already "verified the actual existence of the new planets [meaning the moons of Jupiter] and had been constantly observing them for two months; we compared notes, and I found that their observations agreed exactly with my own."

Stephen Jay Gould on Man's Alleged Self-Importance

"Sigmund Freud often remarked that great revolutions in the history of science have but one common, and ironic, feature: they knock human arrogance off one pedestal after another of our previous conviction about our own self-importance. In Freud's three examples, Copernicus moved our home from center to periphery, Darwin then relegated us to 'descent from an animal world;' and, finally (in one of the least modest statements of intellectual history), Freud himself discovered the unconscious and exploded the myth of a fully rational mind. In this wise and crucial sense, the Darwinian revolution remains woefully incomplete because, even though thinking humanity accepts the fact of evolution, most of us are still unwilling to abandon the comforting view that evolution means (or at least embodies a central principle of) progress."

Stephen Jay Gould, "The Evolution of Life On Earth," *Scientific American*, v. 271, October 1994, 91

With his telescope, Galileo had discovered sunspots, whose visible motion across the sun's surface implied its rotation. But the Jesuits claimed that they had already discovered this, and their subsequent dispute over priority has a very modern ring, suggesting that vanity was more easily offended than theology. Furthermore, Galileo wanted to assert that Copernican theory was true, while the Jesuits insisted that it should be treated as a theory.

Cardinal Bellarmine, a Jesuit and doctor of the Church, wrote to a friend that "you and Galileo would be well advised to speak not in absolute terms but *ex suppositione*, as I am convinced that Copernicus himself did." But Galileo refused, and in 1615 a complaint was lodged with the Inquisition. Again, Galileo was told that a few sentences declaring Copernicus' work to be only theory was all that was needed to clear up the matter.

In 1623 Galileo published *The Assayer*, rejecting in the name of science all authority but observation and reason. It was dedicated to Pope Urban VIII, who sanctioned its publication in 1623 and (we are told) enjoyed it.

Meanwhile, Galileo was working on his important *Dialogue on the Two Chief Systems of the World*. Having shown it to the pope, he was told he could publish it, but only as a theory. Galileo revised the book, which was published in 1632. But he remained headstrong and tactless. In the book's dialogue, the defenders of Copernicanism were Galilieo's close friends, while its opponent was a character, Simplicio, who rejected the new astronomy with childish arguments and transparent sophistries. Toward the book's end, Galileo put into Simplicio's mouth an addition to the text that Pope Urban had explicitly requested.

The insertion was made, but it was imputed to a simpleton. The *Dialogue* was quickly suppressed and unsold copies seized. Galileo was brought back before the Inquisition where, abjuring his heresies, he was sentenced to imprisonment. Within days, the sentence was commuted and he returned to his villa near Florence. He was additionally instructed to recite seven psalms daily for three years. His daughter, a nun, performed this penance for him. Although forbidden to leave the villa's grounds, he was free to pursue his studies, teach pupils, write books, and receive visitors, among them the poet John Milton.

One could conclude that Galileo was looking for trouble and that the church was willing to fight him, but another factor was at play: the Protestant Reformation. Under pressure, the Church was more concerned about Protestant inroads than about science. The church belatedly recognized a need to convince fallen-away Catholics that the Bible was as important to the Holy See as it was to Protestants. In today's terms, one might say that Rome needed to improve relations with the Bible Belt.

Until the Protestant Reformation, scripture was not interpreted so literally. As H. W. Crocker III wrote in *Triumph*, a history of the Catholic Church: "Catholic interpretations of the Bible—such as had come from the Church fathers, like Saint Ambrose and Saint Augustine—had never been literal. Saint Augustine's conversion had even depended on the Bible not having been taken literally."[14] But in the light of the Reformation, scientific speculation at odds with Biblical literalism only promised more headaches.

Furthermore, it was reasonable for the church to insist that Copernican theory had not been proven true. It is only with hindsight that this has acquired "fundamentalist" overtones. Recent scholarship has shown that between the publication of Copernicus's treatise and the year 1600, a period of fifty-seven years, "only ten Copernicans have been identified, in the strong sense of having advocated the earth's physical motion," according to John Brooke.

Among those who had not accepted Copernicus was Tycho Brahe (1546–1601), the Danish astronomer whose data were later used by Johannes Kepler (1571–1630) to formulate the laws of planetary motion. Brahe is regarded as the greatest astronomer of the second half of the sixteenth century. He argued that the planets revolved around the sun, which in turn orbited the earth. If Brahe couldn't accept Copernican theory, was it so unreasonable for the Vatican also to have doubts?

Galileo and Cardinal Bellarmine had actually agreed that there was no need to reinterpret the Bible without demonstrable proof that the earth moves. At the time, proof was lacking, which was why Tycho Brahe didn't accept Copernicanism. It was only later that key evidence was provided, with James Bradley's discovery of the aberration of starlight (a small change in the position of a star due to the earth's motion) in the 1720s, and with the long-sought stellar parallax, finally discovered in the 1830s.

The Catholic Church always has been more open to science than its reputation would suggest. The Vatican opened its own observatory in the sixteenth century and led the way in reforming the calendar, a triumph

Bruno's Bitterness Was Not Scientific

Giordano Bruno, burned at the stake by the Roman Inquisition in 1600, was a Copernican, and "has often been seen as the archetypal scientific martyr," John Brooke has written. But there was no mention of his Copernican views in the record of his trial. Heresies specified against Bruno concerned the Incarnation and the Trinity. Bruno was also a renegade monk, rumored to have declared Christ a rogue, all monks asses, and Catholic doctrine asinine. "Behind his hostility," Brooke added, "lay a conviction that the Roman Church represented a corruption of an earlier undefiled religion that he associated with the Egyptians.... His world-picture was colored by a magical philosophy that almost became his religion." Bruno's interrogators were more concerned about his heretical theology than his advanced astronomy.

of modern astronomy. An earlier revision of the calendar by the Romans, during the time of Julius Caesar, had overestimated the year by eleven minutes. By 1577 the Julian calendar was off by twelve days, lagging behind the seasons, rendering Church feasts anachronistic. In 1582, Pope Gregory XIII established the Gregorian calendar in Catholic countries. Ten days were omitted in October 1582. Protestant countries opposed the change and in some cases did not adopt it until well over a century later: England held out until 1753, and Russia did not adopt it until 1918.

The great goal of science is to discern and describe the laws of nature. Such laws are to be expected within a created cosmos, but

A Book You're Not Supposed to Read

Inventing the Flat Earth by Jeffrey Burton Russell; New York: Praeger, 1991.

not from chaos. If the universe behaved randomly, science could not exist. The philosopher Descartes said he sought to discover the "laws that God has put into nature." Newton declared that the regulation of the solar system presupposed the "dominion of an intelligent and powerful being." In exposing the geometry of creation, Kepler argued, he was thinking God's thoughts after Him.

This is unique to the Judeo-Christian tradition. In Islam, Allah is an arbitrary and willful God, demanding submission. Bow down, pray for good fortune, for at any moment He might reward or punish unpredictably. It would be fruitless and impious to study the handiwork of such a deity. After a creative period following the initial Muslim conquest, during which scholars absorbed much of value from other cultures, Islamic law, government, and science fell into a prolonged and steady decline from which it shows no sign of recovering.

Chapter 13

BY CHANCE, OR BY DESIGN?

No issue arouses so much passion as the claim that the theory of evolution is inadequate to explain life. President Bush stirred things up even more when he said that schoolchildren should be taught about intelligent design. "Both sides ought to be properly taught," he said, "so people can understand what the debate is about."[1]

The proponents of intelligent design say that living organisms are so complex that they could not have been generated by the long series of accidents that Darwinism relies on. All life forms—plants, animals, and human beings—must have been designed.

The Catholic Church has also taken a renewed interest in the subject. Cardinal Schonborn of Vienna wrote an article for the *New York Times* saying that evolution in the sense of an unguided, unplanned, random process is not true. "Scientific theories that try to explain away the appearance of design as the result of 'chance and necessity' are not science at all, but, as [Pope] John Paul put it, an abdication of human intelligence."[2]

Evolutionists oppose any possibility of design. According to one of their leading advocates, Richard Dawkins of Oxford University, animals look designed, but "really are not." On the first page of *The Blind Watchmaker*, he wrote: "Biology is the study of complicated things that give the appearance of having been designed for a purpose." Dawkins is an

Guess what?

.•. Darwin discovered the mechanism of evolution: natural selection. But eminent geneticists and philosophers have said it is a truism.

.•. In the 1950s, Julian Huxley, the grandson of Darwin's greatest defender, saw the need for "true belief" about evolution.

.•. Lehigh's Michael Behe, reviewing the technical literature, found that questions about the origins of complex biochemical systems essential for life had scarcely been addressed, or answered.

atheist, and Darwinism made it possible to be "an intellectually fulfilled atheist," he said.[3]

In the evolutionist worldview, life on earth evolved from inanimate matter over a long period as a result of random events. If it really is true that all creatures great and small appeared on earth in this fashion, then we have no reason to believe that life is anything other than a cosmic accident, purposeless and pointless.

Evolutionists who insist that this doctrine is true must therefore also claim that religion is a delusion. On the other hand, if evolution from the primordial soup by random change is not established by science, but depends instead on the philosophical presuppositions of its supporters, we do need to be told that, and it should be taught in school.

Unsentimental evolutionists like Richard Dawkins, and the leading proponents of intelligent design, such as Phillip E. Johnson, an emeritus professor of law at the University of California in Berkeley, do agree on the following point. The middle ground, wherein evolution is accepted as God's preferred method of creation—theistic evolution, it is sometimes called—doesn't have a great deal to recommend it.

Dawkins is joined in this respect by a number of scientists and philosophers, including Daniel Dennett of Tufts University, the author of *Darwin's Dangerous Idea* (1995). He saw Darwinism as a "universal acid," which "eats through just about every traditional concept and leaves in its wake a revolutionized world."[4] William Provine, an evolutionary biologist at Cornell University, takes a similar view. He has called Darwinism the greatest engine of atheism devised by man.

Phillip Johnson and other advocates of intelligent design accept "microevolution" as the cause of small body changes: variations in moths' coloring or the shape of finches' beaks. But that "doesn't tell you how the moths and birds and trees got there in the first place," Johnson says. "The human body is packed with marvels, eyes and lungs and cells, and evolutionary gradualism can't account for that."[5] Johnson does not believe that these natural wonders arose from the blind, chance-driven processes decreed by evolutionary biology.

Therefore the stakes are high. If Dawkins, Dennett, and Provine are right—and there are many others who agree with them, although they are usually less forthright—God is a fantasy. If the intelligent design proponents are right, then the theory of evolution belongs on the intellectual scrap heap.

The "middle ground" of theistic evolution also has its proponents. They were conspicuous in a 2001 PBS series on evolution that sought to reassure us that "belief in evolution does not challenge religious beliefs." A leading supporter of that view is Kenneth R. Miller, a professor of biology at Brown University and the author of *Finding Darwin's God.* On PBS, Miller was shown attending Catholic Mass, receiving Communion, and preaching good old-fashioned Darwinism for the cameras.

In 2005 Miller was invited to testify in the hearings held by the Kansas State Board of Education. Board members were trying to decide what schoolchildren should be taught about evolution. Convinced that the board had already decided to go ahead and "teach the controversy," Miller declined to appear. It might seem that he was "too afraid to face the intelligent design people in public," he told the *New York Times.* But taking part in such discussions "only contributes to the idea that there is something worth arguing about." He "wasn't interested in playing a role in that."[6]

Richard Dawkins also thinks that evolution is not a debatable topic. "I'm concerned about implying that there is some sort of scientific argument going on," he told *Time* magazine. "There's not." *Time* added, however, that "the strategy of disengagement may be backfiring on those who care about teaching evolution." When they boycotted the discussion of biology standards in Kansas, "they left the floor wide open to critics of evolution, who won the day."[7]

Like almost all evolutionists, Kenneth Miller asserts that "evolution is a fact."[8] How did he arrive at this finding? "It is a fact that we humans did not appear suddenly on this planet, de novo creations without ancestors," he wrote in *Finding Darwin's God*. But that was little more than table-pounding. A fact is a fact is a fact. Authority, in Miller's league, is expected to prevail without too much argument. But he has lots of company, and his many allies will all agree that evolution is a fact.

Is evolution a fact? If defined weakly enough, as "change over time," then obviously it is. But Kenneth Miller goes further, and boldly claims: "It is a fact that the threads of ancestry are clear for us and for hundreds of other species and groups." One would like to see him provide chapter and verse for that claim, but the vague word "threads" may give him a loophole.

Evolutionists say that intelligent design does not rise to the level of a theory, and they may be right. Michael Behe, a professor of biochemistry at Lehigh University and one of the most prominent critics of Darwinism, says "you can't prove intelligent design by an experiment." Behe is a senior fellow of the Discovery Institute, the Seattle think tank where much of the opposition to Darwinian evolution is centered. He is the author of *Darwin's Black Box*, published in 1996. It was the first outright anti-Darwin book published by a major New York house for decades, perhaps since the 1920s.

If the advocates of design can invoke an invisible Designer, or God, who can prevail over all difficulties any time He wants and design any form of life at will, then we are more within the realm of magic than science. If there is nothing that an Intelligent Designer cannot do, then the theory of intelligent design is unfalsifiable, and not scientific for that reason. One critic of intelligent design, Douglas H. Erwin, a paleobiologist at the Smithsonian Institution, told the *New York Times*: "One of the rules of science is, no miracles allowed. That's a fundamental presumption of what we do."[9]

But a comparable criticism can also be leveled at Darwinism. If material causes *only* are admitted, and nothing exists in the universe but molecules in motion, then evolution must be true—a logical deduction from the premise of materialism. We are indubitably here, along with millions of other species, so how did we get here? Materialists have no choice but to accept that the molecules whirled themselves into extraordinarily complex, conscious beings.

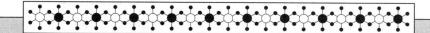

Remember What the Monster Did to Dr. Frankenstein?

Remarkably, evolutionists often use theological arguments to oppose intelligent design. In *Time* magazine, Harvard psychology professor Steven Pinker writes: "Our own bodies are riddled with quirks that no competent engineer would have planned but that disclose a history of trial-and-error tinkering: a retina installed backward, a seminal duct that hooks over the ureter like a garden hose snagged on a tree…"

The designer was incompetent, in other words. No self-respecting deity would use such improvisations. Stephen Jay Gould often made the same argument. In his lengthy diatribe in the *New Republic*, "Unintelligent Design," Jerry Coyne of the University of Chicago relied heavily on presumptuous claims about what an intelligent designer would or would not do.

This is theological criticism masquerading as scientific know-how. The fossil record itself discloses no such "history" of trial and error. That was all in Professor Pinker's imagination. Eyes work perfectly well as they are. Engineers tinkering with cameras haven't been able to produce anything remotely comparable or compact. Microbiologists haven't come up with anything at all. By Darwin's own criteria, our bodies, including the seminal duct, are well designed. If they were not, we wouldn't be here to comment on them.

First produce a better-designed body in your Harvard lab, Dr. Pinker. We'll see what you and your colleagues come up with, then we'll pay attention to your criticisms.

Time, August 15, 2005; *New Republic*, August 22 & 29, 2005

Design theorists today offer a more radical critique than anything seen in Darwin's day. In the nineteenth century, churchmen and the defenders of the traditional order were often awed by science. Religious believers held out for little more than an exemption for the human race—an exception to the evolutionary law. As long as the scientific authorities allowed a "missing link" between ape and man, the clerical establishment was broadly satisfied. Today, however, the advocates of intelligent design challenge any "link" you can think of. It's not that one link is missing, but that none is known for certain.

This rising tide of skepticism has displeased the modern authorities, such as Donald Kennedy, editor in chief of *Science*. He was asked in 1998 if he knew that a majority of U.S. citizens did not believe that humans "descended from other forms." He did know that, he said, but he hoped that things might change. "Well," he said in 2005, "things changed in the wrong direction: alternatives to the teaching of biological evolution are now being debated in no fewer than forty states."[10]

Kennedy blamed a "convergence" of the "present wave of evangelical Christianity" and "partisan loyalty," and added (inaccurately) that "certain kinds of science are now proscribed on what amount to religious grounds," citing stem cell research. (Such research is legal, however. He had confused science itself with the federal funding of science.) Like many evolutionists, he was happy to make a scapegoat of religion. That distracts attention from what may really be at stake: the lack of a credible scientific basis for the dogmatic claims of evolutionists.

The best-known design theorists do not espouse any given religious doctrine, nor do they rely on the Book of Genesis. This has added to the experts' frustration, because they cannot easily dismiss their opponents as fundamentalists. The proponents of design mostly appeal to evidence and reason. Some or most of them are churchgoers, however, and from time to time evolutionists accuse them of making religious claims in the guise of science. The enduring strategy of the evolutionists is to reposition the conflict on the familiar and comfortable terrain of "religion versus science."

Design theorists say that the real, if unacknowledged, support for evolution comes in the form of a philosophy. Thus each side accuses the other of using science as the respectable front for a non-scientific agenda. Design theorists are said to be "creationist" true-believers; evolutionists are said to espouse a worldview—the philosophy of materialism—from which evolution emerges as a logical necessity.

The origin of a theory: Darwinism and its weakness

The idea that life emerged from antecedent life, in forms that become progressively more complex, was suggested long before Charles Darwin published *On the Origin of Species* in 1859. His grandfather Erasmus Darwin was an early evolutionist, and eighteenth-century Enlightenment philosophers were intrigued by the idea of a chain of life going back to inanimate matter. An early evolutionist was naturalist Jean-Baptiste Lamarck, who argued for the inheritance of acquired characteristics. That was shown to be incorrect, but Darwin saw the larger issue at stake. Lamarck had seen that organic change could be "the result of law, and not of miraculous interposition." He praised Lamarck's "eminent service."

It wasn't until Darwin proposed his own evolutionary mechanism, natural selection, that evolutionism began to seem plausible. Naturalists observing organic beings might well conclude that they "had not been independently created," but had descended from "other creatures," Darwin wrote. But such a conclusion would be unpersuasive until it could be shown *how* it happened. How did species "acquire that perfection of structure and coadaptation which most justly excites our admiration"?

Darwin proposed a mechanism. He answered the "how" question, at least to the satisfaction of his peers. His idea was swiftly accepted, and once stated it seemed only too obvious. It is often summarized by the phrase coined by the philosopher of laissez faire, Herbert Spencer—"the survival of the fittest." Darwin himself thought this a good summary of his argument, and added it to later editions of his book.

Some types are fitter than others, the argument ran, and in the competition for survival—the "struggle for existence"—only the fitter variants would survive to propagate their kind. And so animals and plants would become progressively more "adapted" to their surroundings. Those that didn't adapt would die out. Nature itself, then, has "evolving machinery" built into it.

"How extremely stupid not to have thought of that!" said Darwin's great disciple and publicist, Thomas Henry Huxley. He coined the term *agnostic*, and remained one.

By the time of the centennial celebrations at the University of Chicago in 1959, Darwinism was triumphant. Sir Julian Huxley (grandson of Thomas Henry) affirmed that "the evolution of life is no longer a theory,

C. H. Waddington on the Vacuity of Darwinism:

"Everybody has it in the back of his mind that the animals that leave the largest number of offspring are going to be those best adapted also for eating peculiar vegetation, or something of this sort, but this is not explicit in the theory....
There you do come to what is in effect a vacuous statement. Natural selection is that some things leave more offspring than others; and, you ask, which leave more offspring than others; and it is those that leave more offspring, and there is nothing more to it than that. The whole real guts of evolution—which is how do you come to have horses and tigers and things—is outside the mathematical theory."

it is a fact." He added, with more than a trace of dogmatism: "We do not intend to get bogged down in semantics and definitions."[11]

Sir Gavin de Beer of the British Museum remarked that if any layman tried to "impugn" Darwin's conclusions, it must be the result of "ignorance or effrontery." Ecologist Garrett Hardin wrote that anyone who did not honor Darwin "inevitably attracts the speculative psychiatric eye to himself."[12]

Darwin's book had been in print for several decades before biologists began to see just how insubstantial his mechanism was. "For it may appear little more than a truism," wrote eminent Columbia University geneticist Thomas Hunt Morgan, "to state that the individuals that are the best adapted to survive have a better chance of surviving than those not so well adapted to survive."[13] Morgan won the Nobel Prize for his work on the chromosomes of fruit flies.

That problem has never gone away. Logically, no criterion of fitness can be identified that is independent of survival itself. In the end, Darwin's theory of natural selection boils down to the bare claim that some organisms leave more offspring than others. This was acknowledged by British geneticist C. H. Waddington, speaking at the same Darwin centennial:

> Natural selection, which was at first considered a hypothesis that was in need of experimental or observational confirmation, turns out on closer inspection to be a tautology, a statement of an inevitable although previously unrecognized relation. It states that the fittest individuals in a population (defined as those which leave most offspring) will leave most offspring.[14]

Consider the prodigious variety of the millions of sexually reproducing animal and plant species on earth, with their complex adaptations, modes of existence, life cycles, instincts, and means of providing for their young. We learn in school that the great naturalist Charles Darwin discovered the mechanism whereby this great profusion and complexity evolved. Darwin's idea was "the single best idea anybody ever had," said Daniel Dennett, reveling in his hyperbole.

But when that mechanism is more closely analyzed, we find that it amounts to the bare claim that some organisms leave more offspring than others. It surely does leave something to be desired.

The weakness of the Darwinian theory was also recognized by Sir Karl Popper, the preeminent philosopher of science in the twentieth century. The theory is "not testable," but "metaphysical," he wrote:

"To say that a species now living is adapted to its environment is, in fact, almost tautological.... Adaptation or fitness is defined by modern evolutionists as survival value, and can be measured by actual success in survival: there is hardly any possibility of testing a theory as feeble as this."[15]

Any outcome in nature can be regarded as a "confirmation" of Darwin's theory—even the extinction of species. It is sometimes reckoned that 99 percent of all species that ever existed have gone extinct. In that light, Darwinian evolution can be seen as the meager claim that species are well adapted—until they are not. When they fail to adapt, they are "unfit," and cease to exist. So Darwin's theory is once again confirmed. Feeble is the word.

A further weakness of Darwin's theory in explaining the "origin of species" is this. Natural selection does not begin to play any role until self-reproducing organisms already exist. As an explanation for the origin of self-reproducing organisms, therefore, it is a non-starter. The criticism has been made by one of the leading intelligent design theorists, William Dembski, who calls natural selection an oxymoron. For it "attributes the power to choose, which properly belongs only to intelligent agents, to natural causes, which inherently lack the power to choose."[16]

English naturalist Alfred Russel Wallace posited natural selection at about the same time as Darwin. Both authors credited the influence of

Thomas Malthus's *Essay on the Principle of Population*. A population expanding more rapidly than the food supply would create a "pressure" on that population. Both Darwin and Wallace argued that this would tend to eliminate the less fit and promote the more fit.

Malthus was a free-market economist; his *Principles of Political Economy* was published at the same time as Darwin's initial sketch of his theory. British philosopher Bertrand Russell pointed out that Darwin's theory was "essentially an extension to the animal and vegetable world of laissez faire economics."[17]

But only in the twentieth century did these essentially political roots of Darwinism became conspicuous. "Darwin's whole theory of evolution by natural selection bears an uncanny resemblance to the political economic theory of early capitalism," Harvard University geneticist Richard Lewontin has said. "What Darwin did was take early nineteenth-century political economy and extend it to include all of natural economy."[18] Stephen Jay Gould has said much the same thing.

Richard Dawkins, who is as awed as Daniel Dennett was by Darwin's brilliant idea, is nonetheless mystified that "it had to wait until the nineteenth century before anyone thought of it."[19] The reason is simple. Darwinism was mid-Victorian political economy imported into biology.

We are sometimes still inclined to think that the specimen-gathering expeditions of Wallace and Darwin to the Malay Archipelago and the Galapagos Islands (respectively) inspired them. More likely, it was the bustle of competition, the bankruptcies, the poor houses, and the debtors' prisons of Dickensian England. The effect of reading Malthus, Wallace said in a memoir, "was analogous to that of friction upon the specially-prepared match, producing that flash of insight which led us immediately to the simple but universal law of the 'survival of the fittest'."

It is remarkable, surely, that two naturalists had the *same idea* about evolution at the *same time*, after reading the *same book*—a book by an economist.

How Did the Eye Evolve?
David Berlinski vs. Daniel Dennett

Tufts philosophy professor Daniel C. Dennett discussed the evolution of the eye recently in the *New York Times*. It is something that laymen have often found difficult to accept. After a brief discussion of the difficulties—"megabytes of information going into the visual cortex every second for years on end"—Dennett swept them all aside: "But as we learn more and more about the history of the genes involved, and how they work—all the way back to their predecessor genes in the sightless bacteria from which multicelled animals evolved more than a half billion years ago—we can begin to tell the story of how photosensitive spots gradually turned into light-sensitive craters that could detect the rough direction from which light came, and then gradually acquired their lenses, improving their information-gathering capacities all the while. We can't say yet what all the details of this process were, but real eyes representative of all the intermediate stages can be found, dotted around the animal kingdom, and we have detailed computer models to demonstrate that the creative process works just as the theory says."

Mathematician David Berlinski responded:

"It is perfectly plain that Dennett has given up reading the literature. There are eyes throughout the animal kingdom. It's not at all obvious how any of them arose; and still less is it obvious that they arose by any known Darwinian mechanism. To explain the evolution of the eye by appealing to visual systems throughout the animal kingdom is a little like explaining the appearance of *War and Peace* by pointing out that Homer also wrote a war poem, and that Hesiod offered a cosmogeny, and that Virgil also appealed to patriotic sentiments. True enough. But hardly to the point. There is no natural progression that we can trace throughout the paleontological record that begins with a light-sensitive spot and that ends with the eye. If for a moment one allows one's Darwinian faith to lapse, then those so-called intermediates of which Dennett writes so optimistically do not look like intermediates at all. They look like variants against a central type. The Darwinian progression is, of course, entirely an inferential artifact.

"This notion that there is somewhere a computer model of the evolutionary development of the eye is an urban myth. Such a model does not exist. There is no such model anywhere in any laboratory. No one has the faintest idea how to make one. The whole story was fabricated out of thin air by Richard Dawkins. The senior author of the study on which Dawkins based his claim—Dan E. Nilsson—has explicitly rejected the idea that his laboratory has ever produced a computer simulation of the eye's development."

Daniel Dennett, "Show Me the Science," *New York Times*, August 28, 2005; David Berlinski, e-mail to author, August 31, 2005; see also Berlinski, "Has Darwin Met His Match," *Commentary*, December, 2002, and letters, July 2003

But what applies to the world of free-market competition doesn't even begin to apply to the natural world. One company can see what another company is doing, examine the rival's product, and "adapt" accordingly. But the genes of one animal can neither see nor adapt to the genetic changes of another. Mutation is blind, therefore, and the only hope is that the right combination will come up by a roll of the genetic dice. This suggests that Darwin's theory of evolution, far from being "the single best idea anybody ever had," was something far more parochial and culturally generated.

We may question whether the economic insight about competition and survival has any application to the animal and vegetable kingdoms, except to the extent that it is a truism. For we can hardly deny, in Thomas H. Morgan's words, that "the individuals that are the best adapted to survive have a better chance of surviving than those not so well adapted to survive."

Irreducible complexity

In *Darwin's Black Box*, Michael Behe made another important criticism. The question that preoccupied him was: how did complex biochemical systems come into existence in the first place? They are essential for the functioning of life, and their appearance cannot be attributed to chance.

When he researched the technical literature, Behe found that his questions had scarcely ever been addressed, let alone answered. Where there should have been controversy and debate, silence prevailed. He began to see that in graduate school he had assumed that molecular biologists know more about the origin and development of life than they really do.

In fact, they don't have any idea how the mechanisms they study came into existence. A measure of the difficulty of attributing them to chance was provided by Francis Crick, co-discover of the structure of DNA. In 1973, and again in the 1990s, Crick proposed that the earth was "seeded" by spores engineered on a distant planet; "directed panspermia," he called it. He made this proposal only because he knew enough about molecular biology to realize that the undirected origin of life presented tremendous obstacles. To avoid invoking "mind," or the supernatural, he needed to introduce an intelligent designer somehow. Spacemen would do. Crick's Nobel Prize and fame gave him an unusual measure of freedom. Less eminent biochemists usually keep quiet about the difficulties their field presents to the regnant philosophy of materialism.

Behe examined a number of complex biochemical systems in detail: the biochemistry of vision, the blood-clotting system (it "makes a fellow yearn for the simplicity of a cartoon Rube Goldberg machine"), and the cilium, a whip-like device that propels cells through bodily fluids. These structures turn out to have "dozens or even hundreds of precisely tailored parts," Behe wrote. There are thousands of these complex systems, and in not a single case has a plausible mechanism for their origin been offered. So how did they get here?

The answer has not changed since the 1850s. By "numerous, successive, slight modifications," as Darwin put it. One part fortuitously appeared by random mutation, and this conferred a "selective advantage" upon the organism. Then there was another accident, and so on. But this won't work if all the parts have to be present and correct from the beginning. And that is required. Accepting Darwin's explanation is a little like believing that a piston rod will make a car run a little bit, and then, if you connect it to a crank shaft, it will run a little bit better. Finally, when all the parts are in place, it will get thirty miles to the gallon.

Behe illustrated the argument with a mousetrap. It's simple enough, but all parts must be properly aligned before the trap can catch one mouse. It is "irreducibly complex," as Behe put it. In the *Origin of Species*, Darwin took a forlorn stab at explaining vision by imagining at the outset a light-sensitive spot. This conferred a survival advantage and so more offspring inherited the structure. Richard Dawkins has since repeated the claim. The problem is that a minimally functioning system of vision must begin with an array of cells that "make the complexity of a motorcycle or a television set look paltry by comparison," Behe wrote.

Darwin did not know this. But David Berlinksi, who has written in *Commentary* about the problems associated with the evolution of the eye, says that Dawkins was essentially making it up. Daniel Dennett addressed the evolution of the eye in a recent article for the *New York Times*. Berlinski then criticized Dennett, and Dawkins for good measure.

Until recently, evolutionists could take refuge in ignorance. Structures at the molecular level were not known. Scientists could make the convenient assumption that the organization of matter at the submicroscopic scale was straightforward. All the real problems of evolution could be relegated to a "black box," as Behe put it, which no one could inspect. Like children's plastic toys, simple parts were visualized as easily interlocking. Insects, it was believed, arose from spoiled food. Ernst Haeckel, an early admirer of Darwin, assured his readers that the cell itself was "a simple little lump of albuminous carbon."[20]

Now we know better. Too many improbable events would have to occur at the same time for a chain of accidents to cause life. As to the explanations that have been attempted, Behe said that 80 percent of the articles in the *Journal of Molecular Evolution* compare amino-acid sequences of proteins from different species, and are content to identify similarities and dissimilarities in these sequences. But sequence comparisons can no more tell us how a complex system arose than comparable passages from two computer manuals can tell us how the computers themselves were assembled, or whether a computer can be assembled step-by-step starting from a typewriter. It can't. You have to start over from scratch.

Books You're Not Supposed to Read

Uncommon Dissent: Intellectuals Who Find Darwinism Unconvincing by William A. Dembski, ed.; Wilmington, DE: ISI Books, 2004.

Darwin's Black Box: The Biochemical Challenge to Evolution by Michael J. Behe; New York: Free Press, 1996.

From Darwin to Hitler, Evolutionary Ethics, Eugenics and Racism in Germany by Richard Weikhart; New York: Palgrave Macmillan, 2004.

But so far we have looked only at Darwin's mechanism, natural selection. What about the evidence that evolution actually has occurred? How strong is that? David Berlinski once wrote that evolution is a process that "has not been observed." Facts in favor of it "have been rather less forthcoming than evolutionary biologists might have hoped," he wrote. So let's take a closer look.

Chapter 14

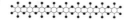

EVOLUTION
The Missing Evidence

What is the evidence that evolution did in fact take place? "Evolution" here must mean something more than "change over time." Dinosaurs once existed, and they no longer do. So the animal kingdom surely has changed over time. But that is not what we mean by evolution.

We know that in the aggregate, the genetic composition of a population will change over time. But that is not what we mean either. We know that the melanic, or dark-colored, percentage of certain species of moth increased, relative to the "speckled" or light-colored population, in parts of industrial England in the twentieth century. Do such examples show that evolution is a fact? Surely not. Changes in gene ratios are indubitable, omnipresent, and trivial. Evolution implies something far more than that.

Geneticist Thomas H. Morgan said a hundred years ago that "evolution means making new things, not more of what already exists." Both "melanic" and "speckled" varieties of moth already existed before their relative percentages changed. Evolution entails the emergence of a new type of moth.

For the layman, evolution implies that one species is connected by an ancestral chain to another. Or it means that a group of organisms that are undeniably related, such as bats, bears and whales—all are mammals—share their common features *because* all are descended from the same

Guess what?

.ɪ̇. A senior paleontologist from the British Museum challenged experts to tell him "anything you know about evolution... that you think is true?" The only answer he got was silence.

.ɪ̇. The oldest fossil bats already have echolocation, or sonar, built in.

.ɪ̇. In *On the Origin of Species*, Charles Darwin relied on sketches that turned out to be doctored.

215

ancestral mammal. If evolutionary biologists wish to tell us that evolution is a fact, then we expect to be given evidence for claims such as that.

One of the most remarkable discussions of what we know about evolution was held at the American Museum of Natural History, on Central Park West in New York City. The speaker was Colin Patterson, a senior paleontologist at the British Museum, who was then visiting the United States. In November 1981, he addressed the Systematics Discussion Group, consisting mostly of professional biologists and museum staffers with a particular interest in animal classification. They would meet once a month in a classroom opposite the dinosaur exhibit of the Natural History Museum.

Patterson had already caused a controversy three years earlier, saying in a pamphlet published by the British Museum: "If the theory of evolution is true..." That set off weeks of agitation and a flurry of letters to *Nature*. Patterson always emphasized that he had no religious agenda—religion was all a "pack of lies," he once said. What he most strongly opposed was the confusion of knowledge and faith. A faithful Darwinian who was stunned by his skeptical outlook once asked him if he "believed in" evolution. Patterson said that he did, but he added that scientific claims weren't supposed to be matters of faith.

The transcript of Patterson's talk was later reviewed and corrected by his friend Gary Nelson, who had been present that day. Nelson was then chairman of the ichthyology department at the Natural History Museum. Here is how Patterson began:

> Now, I think always before in my life, when I've got up to speak on a subject, I've been confident of one thing—that I know more about it than anybody in the room, because I've worked on it.
>
> Well, this time that isn't true. I'm speaking on two subjects, evolutionism and creationism, and I believe it's true to say that

I know nothing whatever about either of them. Now, one of the reasons I started taking this anti-evolutionary view, well, let's call it non-evolutionary, was last year I had a sudden realization. For over twenty years I had thought that I was working on evolution in some way. One morning I woke up, and something had happened in the night, and it struck me that I had been working on this stuff for twenty years, and there was not one thing I knew about it. That was quite a shock, to learn that one can be so misled for so long.

So either there was something wrong with me, or there was something wrong with evolutionary theory. Naturally I know there's nothing wrong with me. So for the last few weeks, I've tried putting a simple question to various people and groups of people.

The question is: can you tell me anything you know about evolution, any one thing, any one thing that you think is true? I tried that question on the geology staff in the Field Museum of Natural History, and the only answer I got was silence. I tried it on the members of the Evolutionary Morphology Seminar in the University of Chicago, a very prestigious body of evolutionists, and all I got there was silence for a long time, and then eventually one person said, "Yes, I do know one thing. It ought not to be taught in high school." [Laughter]

Patterson knew many people in the audience, and some were friends. There were elements of hyperbole and humor in his remarks. But no one was prepared to argue when he added, a minute or two later, that "it does seem that the level of knowledge about evolution is remarkably shallow."

By this time, he had already written an introductory text called *Evolution*, published by the British Museum. After it came out, a curious reader wrote to him and asked why he had not included in the book any "direct illustrations of evolutionary transitions." Patterson replied:

> You say I should at least "show a photo of the fossil from which each type of organism was derived." I will lay it on the line—there is not one such fossil for which one could make a watertight argument. The reason is that statements about ancestry and descent are not applicable in the fossil record. Is *Archaeopteryx* the ancestor of all birds? Perhaps yes, perhaps no: there is no way of answering the question. It is easy enough to make up stories of how one form gave rise to another, and to find reasons why the stages should be favoured by natural selection. But such stories are not part of science, for there is no way of putting them to the test.[1]

The recording of Patterson's talk was made by a member of the audience without his knowledge, and he later said that it had sometimes been quoted inaccurately. But he never repudiated his remarks, and in fact revisited them in a second talk, in London in 1993. This time he seemed to cast even greater doubt about what we know—at least about molecular data and whether that tells us anything abut evolution. In any event, a transcript of the talk is available at the Access Research Network, online, and the curious can even obtain an audio copy, either on CD or cassette.

"There are no half-bats."

Colin Patterson, who died in 1998, belonged to the school of taxonomists called transformed or "pattern" cladists. Their main argument is that all we see in the fossil record are patterns (of similarity or difference) and in themselves these patterns do not tell us how they arose. In short, we cannot deduce a *process* from a *pattern*. The founding father of cladistics was a German entomologist named Willi Hennig, who made many original observations and discoveries in the field of systematics (a study of the interconnections between various groups).

The terminology of cladistics is often obscure, but one of Hennig's most important observations can be easily understood. It had also been made by Aristotle. Many groups are defined by an absence of characteristics, Hennig said, and these are not proper groups. The best known is the group *invertebrates*. Absolutely anything qualifies as long as it is not a vertebrate.

Pondering this point, and studying the fossil data and the claims made about fossils by evolutionists, Patterson then made a remarkable claim. All the well-known ancestral groups of evolutionary biology are of this type, he said—they are all defined by an *absence* of characteristics. And statements claiming to identify such groups as ancestral to other groups are disguised tautologies, Patterson said. They are true by definition.

Consider the claim "vertebrates evolved from invertebrates." It is merely a roundabout way of saying that the ancestor of the first vertebrate was not a vertebrate—which is true by definition—otherwise the "first" vertebrate would not be the first. "Cats evolved from non-cats" is a comparable statement, and if you think about it for a few seconds, you see that it conveys no real knowledge. A logical relationship is dressed up as an empirical claim. Something that was "observed" in the dictionary is made to seem as though it was observed in the rocks. Which it wasn't.

Evolutionists believe that they do know the process that creates similarities, sometimes known as homologies. If two otherwise dissimilar animals have identifiable backbones, for example, it is said that they share this feature because they also shared a common ancestor from whom they inherited that common trait.

The similarity between structures in different species is often so great that no one can believe that it is accidental—the forelimbs of mammals, for example. The shared features of the bat's wing, the porpoise's fin, and the human hand are so striking that they are unmistakably the "same" thing, although differing in size and proportion. How did that similarity arise? That is the most basic question in evolutionary biology. There must be a cause.

Before Darwin, and before the acceptance of evolution, anatomists such as Richard Owen attributed homologies to a shared "archetype." This was construed variously as a disembodied Platonic idea, or a plan in the mind of the creator. It implied design and intelligent causation.

But Darwin construed homology as evidence for common descent. There was an original creature with this characteristic forelimb, and over many generations its offspring were slowly transformed into bats, porpoises, or humans while retaining the same basic body plan. This purely naturalistic explanation rendered all others superfluous. No longer did we have to entertain notions of archetypes, design, or designers.

In short, Darwin took homologous structures as evidence for evolution. It wasn't the only category of evidence that he offered in the *Origin of Species*, but it was an important part of his argument. (His whole book, he said, was "one long argument.")

More recently, however, there has been a subtle shift. Ernst Mayr, perhaps the dominant evolutionary biologist of the twentieth century, decided that the time had come to *define* homology as a feature found in two or more groups that were "derived from the same (or a corresponding) feature of their common ancestor."[2]

The change here may seem trivial, but notice Mayr's sleight of hand. What Darwin proposed as the *explanation* for homology had now become its definition. This smuggled in the assumption that we have a way of identifying common ancestors other than just looking at fossil bones. Mayr's argument implied that we possess a family tree of interconnecting species—something equivalent to an enormous chart on the wall covering millions of years and enabling us to look up common ancestors at will.

But we have no such chart. All we have are bones scattered in the mud. Relying on fossils, as we must, we have no way of identifying common ancestors other than by contemplating homologous structures.

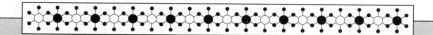

"Backwoods USA"

One of the great mysteries in the history of life is the so-called Cambrian explosion, about 530 million years ago. Perhaps thirty-five out of forty total phyla, or animal body plans, appeared on Earth, all within a very short time, and they have no clear antecedents in the rocks.

Steven Meyer of the Discovery Institute wrote an article about this, published by the peer-reviewed *Proceedings of the Biological Society of Washington*. He relied on the work of scientists at Yale, Oxford, and elsewhere. But Richard Sternberg, the editor of the journal, was immediately in hot water with his colleagues for publishing the article. The chairman of the zoology department called Sternberg's supervisor and asked whether Sternberg was a religious fundamentalist. Was he affiliated with any religious organization? Was he a right-winger?

Sternberg had to surrender his office and keys to the department floor, denying him access to the specimen collections he needed. David Klinghoffer wrote in the *Wall Street Journal*:

> The Biological Society of Washington released a vaguely ecclesiastical statement regretting its association with the article. It did not address its arguments but denied its orthodoxy, citing a resolution of the American Association for the Advancement of Science that defined Intelligent Design as, by its very nature, unscientific.

A senior Smithsonian scientist complained that publication of the article "made us into the laughing stock of the world, even if this kind of rubbish sells well in backwoods USA."

Notice, it was not the substantive claims about the Cambrian Explosion that caused such fury, it was their publication in a peer-reviewed journal.

David Klinghoffer, *Wall Street Journal*, January 28, 2005; Michael Powell, *Washington Post*, August 19, 2005

Mayr's triumph was to have insinuated what he wanted to believe and dedicated his life to promoting: the idea that evolution has already been established as a fact. (A professor of zoology at Harvard for many years, Mayr died in 2005 at the age of one hundred.)

The problem with arguing that similarity of structure is evidence for evolution is this: there are some remarkable similarities of structure that not even Darwinian biologists attribute to common descent. "The structure of an octopus eye is remarkably similar to the structure of a human eye," Jonathan Wells wrote, "yet biologists do not think that the common ancestor of octopuses and humans possessed such an eye."[3]

Even when there is a congruent pattern of similarities in different groups, as in the forelimbs of bat, porpoises, and human, and biologists attribute that similarity to common descent, they are guessing. Not only do we do not possess the unbroken chain of fossils leading back to that shared ancestor, but we hardly have any links in the chain. That is why Patterson said that "statements about ancestry and descent are not applicable in the fossil record."

The earliest fossils of particular species often have a way of appearing suddenly, as though they sprang into existence fully formed. Bats, which come in 1,100 living species (20 percent of all mammal species), are the only mammals capable of powered flight, and yet the oldest bat fossils already have sonar built in. Both sonar and flight arose at about the same time, and quite suddenly. "The lineage leading to bats was thus characterized by two remarkable specializations seen in no other land mammals," according to a recent analysis in *Science*.[4]

Sonar is an extraordinarily sophisticated device involving simultaneous adaptations of the ear, brain, musculature, and respiratory systems. You would have thought that if it emerged as the accumulation of many accidental steps, each one beneficial to the bat-in-progress, as Darwinism decrees, then the fossil terrain would include half-bats, near-bats, almost-made-it bats; would-have-avoided-that-cliff-with-better-sonar bats, and so on. But we never find them. "There are no half-bats," as J. D. Smith, a leading expert on bats, once told a conference of professional biologists.

Few animals are preserved, to be sure, but those that were handicapped compared to their better-adapted successors (to adopt Darwin's storyline) would surely be more likely than their "new and improved" competitors to have stumbled into swamps and tar pits and so have been preserved.

Henry Gee, the author of *In Search of Deep Time* and a *Nature* editor (he worked as an assistant to Colin Patterson for a while), wrote that "the intervals of time that separate fossils are so huge that we cannot say anything definite about their possible connection through ancestry and descent." Each fossil is "an isolated point," he added, with "no knowable connection to any other given fossil." All are immersed "in an overwhelming sea of gaps."[5]

All the physical evidence for human evolution, enough to fill ten thousand banner headlines and newsmagazine covers, "can be put, with room to spare, into a single coffin," according to Lyall Watson. ("A small box," says Gee.) In one account, there are "no fossil chimpanzees," in another, no fossilized "chimpanzee skulls." All is guesswork. It's hardly worth reading about last year's femur fragment found in the Serengeti, because next year there will be a new fragment, new headlines, and new family trees solemnly redrawn on the inside page. All such trees share this feature. They will identify species at the outer tips of the tree—extant or living species such as monkeys or lemurs—but will decline to locate *any identifiable species* at any of the branching points within the tree.

One of the leading advocates of intelligent design is Jonathan Wells, a Discovery Institute fellow with a doctorate in molecular biology from Berkeley. "Even if the fossil record were complete, and it preserved all the desired characters, it would not establish that homology is due to common

A Day at the Griffith Planetarium in the 1950s

LECTURER (O.S.) And while the flash of our beginning has not yet traveled the light years into distance—

Full shot. The dome. The star rushes nearer, looming larger and larger. The music rises in tension and volume.

LECTURER (O.S.) Has not yet been seen by planets deep within the other galaxies, we will disappear into the blackness of the space from which we came.

Two shot. JIM and PLATO staring upward, cringing back into their seats as the light on their faces increases. Music is up loud.

Full shot. The dome seen past PLATO's head. The heavens grow brighter as the star plummets near. Music at crescendo.

LECTURER (O.S.) Destroyed as we began in a burst of gas and fire.

The sky is blasted by a wild flash of light. Music reaches explosion. The stars appear again.

Moving shot. Faces of normal kids watching seriously—very impressed.

LECTURER (O.S.) (continuing) The heavens are still and cold once more. In all the complexity of our universe and the galaxies beyond, the Earth will not be missed.

Medium shot. JIM and PLATO looking up.

LECTURER (O.S.) Through the infinite reaches of space, the problems of Man seem trivial and naive indeed. And Man, existing alone, seems to be an episode of little consequences.

PLATO ducks his head down on the back of JIM's chair. JIM looks at him.

LECTURER (O.S.) That's all. Thank you very much.

The lights go on. The rustle and confusion of kids stretching after sitting too long. Scattered applause. JIM rises and ruffles PLATO's hair.

JIM Hey, it's over. The world ended.

PLATO looks up at him.

PLATO What does he know about Man alone?

From *Rebel Without a Cause:* The scene from the high school field trip to the Griffith Planetarium.

ancestry," he wrote in *Icons of Evolution*. The problem was inadvertently illustrated by Tim Berra, a professor of zoology at Ohio State.

In a 1990 book defending Darwinian evolution against its critics, Berra compared the fossil record to a series of automobile models. "Everything evolves, in the sense of 'descent with modification,' whether it be government policy, sports cars, or organisms," Berra wrote. His argument went like this:

> If you compare a 1953 and a 1954 Corvette, side by side, then a 1954 and a 1955 model, and so on, the descent with modification is overwhelmingly obvious. This is what [paleontologists] do with fossils, *and the evidence is so solid and comprehensive that it cannot be denied by reasonable people*."[6] [Emphasis in original.]

This only shows the extent to which evolutionists find it difficult to think outside their own particular box. As Jonathan Wells noted, Berra "actually spotlights the problem of using a sequence of similarities as evidence for Darwin's theory. We all know that automobiles are manufactured according to archetypes (in this case plans drawn up by engineers), so it is clear there can be other explanations for a sequence of similarities besides descent with modification."[7]

Phillip E. Johnson identified this as "Berra's Blunder." The Corvette sequence, Johnson wrote, "does not illustrate naturalistic evolution at all. It illustrates how intelligent designers will typically achieve their purpose by adding variations to a basic design plan. Above all, such sequences have no tendency whatever to support the claim that there is no need for a Creator....On the contrary, they show that what biologists present as proof of 'evolution' or 'common ancestry' is just as likely to be evidence of common design."[8]

Wells's *Icons of Evolution: Science or Myth?*, like Behe's book, was a landmark in the intelligent-design movement. Instead of simply pointing to difficulties, Wells played offense. The book's subtitle was "Why Much of What We Teach about Evolution Is Wrong." Here are several examples given by Wells, showing where evidence was invented or misrepresented, or where contradictory evidence is suppressed.

Haeckel's embryos

Darwin thought that "by far the strongest single class of facts" in favor of his theory came from embryology. He relied on German biologist Ernst Haeckel, whose drawings of embryos from various classes of vertebrates showed them to be virtually identical in their earliest stages. They become noticeably different only as they developed. This was the pattern that Darwin found so convincing.

Biologists have known for over a century that vertebrate embryos in fact never do look as similar as Haeckel drew them. It turned out that in some cases, Haeckel simply used the same woodcut for embryos that were then represented as belonging to different classes. In other cases, he doctored his drawings to make the embryos appear more alike than they really were. Haeckel's own contemporaries were critical of his work, and charges of fraud abounded in his lifetime.

In 1997, British embryologist Michael Richardson and an international team compared Haeckel's drawings with photographs of actual vertebrate embryos, demonstrating conclusively that the drawings misrepresented the truth. Richardson was quoted in *Science*: "It looks like it's turning out to be one of the most famous fakes in biology."[9]

Yet some version of Haeckel's drawings could be found in most current biology textbooks when Wells's book came out (and possibly still can today). Stephen Jay Gould wrote that we should be "astonished and ashamed by the century of mindless recycling that has led to the persistence of these drawings in a large number, if not a majority, of modern textbooks."[10]

Peppered moths

Darwin had no direct evidence of natural selection when he wrote the *Origin of Species*, so he gave imaginary illustrations. Then, in the 1950s, Bernard Kettlewell seemed to find conclusive evidence of natural selection in Britain. During the previous century, most peppered moths in England had shifted from being light-colored to being dark-colored. It was thought that the dark coloring gave them better camouflage on pollution-darkened tree trunks, protecting the darker moths from predatory birds.

To test this, Kettlewell released light and dark moths onto nearby tree trunks in polluted and unpolluted woodlands, then watched as birds ate the more conspicuous moths. As expected, they caught more light moths in the polluted woodland, and more dark moths in the unpolluted one. In *Scientific American*, Kettlewell called this "Darwin's missing evidence." Peppered moths soon became the best example of natural selection in action, and the story was retold in biology textbooks, illustrated by photographs of the moths on tree trunks.

In the 1980s, however, researchers found that peppered moths don't normally rest on tree trunks. They fly at night and apparently hide under branches during the day. By releasing them onto tree trunks in daylight, Kettlewell had created an artificial situation, and many biologists now consider his results invalid. As for the photos of moths on tree trunks, they were all staged. Photographers even glued dead moths to trees. The people who staged them thought they were representing the true situation, but they were mistaken. Yet they are still used as evidence for natural selection in current biology textbooks.[11]

The tree of life

If all living things are gradually modified descendants of one or a few original forms, Darwinism predicts that the history of life should resemble a branching tree. But this has turned out to be wrong in important ways. The fossil record shows the major groups of animals appearing fully formed at about the same time in a "Cambrian explosion," rather than diverging from a common ancestor. Darwin knew this, and considered it a serious objection to his theory. But he attributed it to the imperfection of the fossil record and believed that future research would supply the missing ancestors.

But almost 150 years of fossil collecting has made the problem worse. Instead of slight differences appearing first, the greatest differences appear right at the start. Some fossil experts note that this "top-down evolution," contradicts the "bottom-up" pattern predicted by Darwin's theory. Yet most biology textbooks don't even mention the Cambrian explosion, much less point out the challenge it poses for Darwinian evolution.

Canadian molecular biologist W. Ford Doolittle doesn't think the problem will go away, and speculated in 1999 that scientists "have failed to find the 'true tree'." Nevertheless, biology textbooks continue to assure students that Darwin's Tree of Life is a scientific fact overwhelmingly confirmed by evidence. Judging from the real fossil and molecular evidence, however, it is a hypothesis masquerading as a fact.[12]

"Building blocks"...in a flask

In 1953 it was widely reported that scientists Stanley Miller and Harold Urey had succeeded in creating "the building blocks of life" in a flask. Mimicking what were believed to be the natural conditions of Earth's early atmosphere, and sending an electric spark through the mixture, Miller and Urey had formed simple amino acids. As they are the "building blocks" of proteins, and proteins are the "building blocks" of life, it was thought that scientists might soon create living organisms.

It appeared to be a dramatic confirmation of evolution. Life wasn't a "miracle" after all. No outside agent or divine intelligence was necessary.

Put the right gases together, add a jolt of electricity, and life was bound to happen. Carl Sagan could confidently predict on television that the planets orbiting those "billions and billions" of stars out there must be teeming with life.

There were problems, however. Scientists were never able to get beyond the simplest amino acids in their simulations, and the creation of proteins began to seem not a small step or a few steps. It involved a great, perhaps impassable divide. An amino acid is to a living organism what a letter of the alphabet is to a Shakespearean play.

Then, in the 1970s, scientists began to believe that the Earth's early atmosphere was nothing like the mixture of gases used by Miller and Urey. Instead of being a hydrogen-rich environment, it probably consisted of gases released by volcanoes. But put those gases in the Miller-Urey apparatus, and the experiment doesn't work at all.

Nonetheless, textbooks continue to use the Miller-Urey experiment to argue that scientists have demonstrated an important first step in the origin of life. This includes *The Molecular Biology of the Cell*, co-authored by the National Academy of Sciences president, Bruce Alberts. They omit to say that the researchers themselves now acknowledge that an understanding of the origin of life still eludes them.[13]

Darwin's finches

A quarter of a century before Darwin published the *Origin of Species*, he was formulating his ideas as a naturalist aboard the British survey ship HMS *Beagle*. When the *Beagle* visited the Galapagos Islands in 1835, Darwin collected specimens of the local wildlife, including some finches.

Though the finches had very little to do with Darwin's development of his theory, they have attracted considerable attention from modern biologists as further evidence of natural selection. In the 1970s, Peter and

Fudging Facts on Finches

"A few years later there was a flood [in the Galapagos], and after that the beak size went back to normal. Nothing new had appeared, and there was no directional change of any kind. Nonetheless, that is the most impressive example of natural selection at work that the Darwinists have been able to find after nearly a century and a half of searching.

"To make the story look better, the National Academy of Sciences removed some facts in its 1998 booklet on *Teaching About Evolution and the Nature of Science*. This version omits the flood year return-to-normal and encourages teachers to speculate that a 'new species of finch' might arise in two hundred years if the initial trend towards increased beak size continued indefinitely. When our leading scientists have to resort to the sort of distortion that would land a stock promoter in jail, you know they are in trouble."

Phillip Johnson, "The Church of Darwin," *Wall Street Journal*, August 16, 1999

Rosemary Grant noted a 5 percent increase in beak size after a severe drought, because the finches were left with only hard-to-crack seeds. The change, though significant, was small; yet some Darwinists claim it explains how finch species originated in the first place.

A 1999 booklet published by the National Academy of Sciences describes Darwin's finches as "a particularly compelling example" of the origin of species. Citing the Grants' work, the booklet explains how "a single year of drought on the islands can drive evolutionary changes in the finches." It calculated that "if droughts occur about once every ten years on the islands, a new species of finch might arise in only about two hundred years." But it failed to point out that the finches' beaks returned to normal when the rains returned. No net evolution occurred. In fact, several of these finch species now appear to be merging through hybridization, rather than diverging through natural selection as Darwin's theory requires.

Withholding evidence in order to give the impression that Darwin's finches confirm evolutionary theory bordered on scientific misconduct. As Phillip Johnson wrote in the *Wall Street Journal* in 1999:

"When our leading scientists have to resort to the sort of distortion that would land a stock promoter in jail, you know they are in trouble."[14]

"On the tendency of varieties to depart indefinitely from the original type"

In a letter he sent to Darwin in 1858, Alfred Russel Wallace proposed something very similar to Darwin's own theory. Wallace's paper was presented at the Linnean Society later that year, along with earlier writings by Darwin, establishing co-discovery. Wallace's contribution was titled "On the tendency of varieties to depart indefinitely from the original type."

This alleged tendency is indeed implied by the theory of evolution, and "indefinite departure" is a prediction of the theory. If evolution from primordial soup to today's living animals is true, then indefinite departure from the original type must have happened. Evolutionists have not been able to demonstrate it in the lab—but not for lack of trying.

Both Wallace and Darwin made their case for natural selection by analogy with the experiments of animal breeders. Darwin bred pigeons himself and spent time with animal fanciers. They were interested in developing certain traits (length of feather, thickness of wool), and they noticed that the offspring of a selected pair often had the trait more abundantly than its parents.

But the breeders reported that if you kept on trying to "push" the animal by selecting repeatedly for certain desired traits, offspring tended to revert to the mean after a few generations. There was an envelope of variability beyond which it was difficult, and perhaps impossible, to push the species. They could be moved around on a plateau—a sizeable plateau in the case of dogs—but not off it altogether.

Wallace responded to this objection by saying that bred animals were pampered and therefore didn't have to struggle. Farmers and breeders fed them and protected them. In the wild, on the other hand, confronted with the "struggle for existence," animals had to exercise all their wiles. Darwin responded differently—with the following rhetorical flight: "How fleeting are the wishes and efforts of man! How short his time! And consequently how poor will his productions be, compared with those accumulated by nature during whole geological periods."[15]

Breeders just put animals together in a pen and said get on with it. Nature, on the other hand, was "daily and hourly scrutinizing," "silently and insensibly working . . . at the improvement of each organic being."

And so *natural* selection could do what *human* selection had been unable to demonstrate. In short, the experience of breeders was countered with rhetoric, analogy, and extrapolation. Darwin argued in the first edition of the *Origin* that, given enough time, it wasn't too much of a stretch to claim that bears could turn into whales.

Experiments began on fruit flies about a hundred years ago, and they have been continuing ever since. With a two-week life cycle, fruit flies are an ideal experimental animal. Females produce literally hundreds of eggs, and the animal rights people don't object. The fruit fly's genome has already been "decoded"—and it turns out to have half as many genes as humans.

Fruit flies have been the chosen instrument for studying the effects of selection pressure over hundreds of generations and tens of thousands of experiments. Temperature and many other environmental factors have been varied. In 1926 geneticist Hermann J. Muller made the famous discovery that X-rays cause genes to mutate. For a number of years, zapping fruit flies with X-rays was assumed to be the most promising way of getting them to evolve into something else—something that was not a fruit fly. Hopes were high.

Books You're Not Supposed to Read

The Design Inference: Eliminating Chance Through Small Probabilities by William A. Dembski; Cambridge: Cambridge University Press, 1998.

The Wedge of Truth: Splitting the Foundations of Naturalism by Phillip E. Johnson; Downers Grove, IL: InterVarsity Press, 2000.

"New Discovery Speeds Up Evolution," *Scientific American* reported in 1928. "Professor Muller's experiments signify [that] evolutionary changes, or mutations, can be produced 150 times as fast by the use of X-rays as they can by the ordinary processes of nature."[16]

So they were well on their way, or so it seemed. But most of the flies were killed outright, and the offspring that maybe showed some "incipient" speciation just didn't seem to want to play the game. Muller produced an "eyeless" fruit fly, but ten generations later, its descendants were found to have reverted to normal. The eyes were back! Muller received the Nobel Prize anyway, in 1946.

Meanwhile, we continued to hear from the breeders and horticulturalists, the most famous of whom was Luther Burbank. He spent the better part of fifty years hybridizing fruits and plants in Santa Rosa, California. He observed a quite different "law"—a law of reversion to the mean. Its empiricism stands in sharp contrast to the evolutionists' theoretical but not yet observed "indefinite departure."

> I know from my experience that I can develop a plum half an inch long or one 2.5 inches long, with every possible length in between, but I am willing to admit that it is hopeless to try to get a plum the size of a small pea, or one as big as a grapefruit. I have daisies on my farms little larger than my fingernail and some that measure six inches across, but I have none as big as a sunflower, and never expect to have. I have roses that bloom pretty steadily for six months in the year, but I have none that will bloom twelve, and I will not have. In short, there are limits to the development possible, and these limits follow a law. But what law, and why?
>
> Experiments carried on extensively have given us scientific proof of what we had already guessed by observation; namely, that plants and animals all tend to revert, in successive

generations, toward a given mean or average.... There is undoubtedly a pull toward the mean which keeps all living things within some more or less fixed limitations.[17]

No Nobel for Luther Burbank! He was non-religious, too, even publishing a pamphlet titled "Why I Am an Infidel," for which he was much reviled. He surely was a scientist, though, and today, after a hundred years of fruitless fruit-fly experiments, it seems that his Law of Reversion to the Mean may have been rather better confirmed than the hoped-for Tendency of Varieties to Depart from the Original Type.

Breeders are still searching for an (undyed) blue rose and black tulip, the coveted but still elusive prizes of horticulture.

Discussing the prolonged failure of the Darwinians to demonstrate speciation, Jonathan Wells said:

> Despite many heroic experiments over the past forty years, the best anyone has ever done is to produce partial or temporary reproductive isolation. After decades of trying unsuccessfully to find evidence for neo-Darwinian speciation, neo-Darwinists have now come to the conclusion that one should not expect to see such evidence because speciation takes too long.

He added the following comment:

> Darwinists claim that their theory is so thoroughly confirmed by overwhelming evidence that it is an uncontroversial "fact" that deserves a complete monopoly in biology and in science classrooms. Yet everything in Darwinian evolution hangs on the origin of species: how can Darwinian evolution account for the evolution of mammals from amoebas, if it can't even account for the origin of one fruit fly species from another fruit fly species? The origin of species is the starting point for all the grandiose claims of "descent with modification," including

universal common ancestry and the creative power of natural selection. This is why Darwin called his magnum opus *On the Origin of Species*—not "How Existing Species Change Over Time."[18]

They used to talk about "the missing link." Today, it's not clear that we have any links. They pushed and pulled, they turned up the heat and turned it down again, they zapped fruit flies for a hundred years. They turned on the X-ray machines and they wheeled in the computers; they spelled out the critter's genome, base by base. But the fruit flies and their offspring, the ones that survived the heat and the X-rays, are still floating around in the lab, feeding as always on rotten fruit, and waiting for something more interesting to turn up. The scientists are waiting for a miracle, too, telling themselves that evolutionism is a fact, enjoying their monopoly and telling everyone that creationism shouldn't be taught in high school—because it's religion, not science.

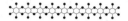

FINAL THOUGHTS

One reason that science has become so politicized is that the federal government transformed itself from a government of limited and specified powers to an all-purpose caring agency. Once upon a time, it provided for the common defense and a common currency. Then the restraints gave way, like the New Orleans levees, and it took on any role that could be called compassionate. Soon it was awash in a flood of issues and missions, and it became less and less able to cope with any of them.

Science hopped on board. If the discovery of emergencies and crises entitled you to a share of federal largesse, scientists could play that game. They had the equipment, after all, the measuring devices, the radar, the thermometers, the satellite sensors.

A chicken in the Orkneys died of a mysterious ailment? A fowl epidemic might be heading our way! Seven people came down with a strange flu in Ho Chi Minh City? Call Lawrence Altman at the *New York Times*! Eight drops of mercury were found in a Washington, D.C., high-school basement? More surveillance required!

As recently as 1989, the budget of the National Institutes of Health was $7.9 billion. By 2005 it had almost quadrupled to $28.8 billion. In his odd but interesting book, *Science, Money, and Politics*, Daniel S. Greenberg, who for years published a newsletter about science and politics, said:

NIH was not a hard sell [in Congress]. Faith in the great scientific center of disease fighting was a non-ideological, bipartisan verity of Capitol Hill. Political support arose naturally, from fear and hope, but was also cultivated by the NIH management.

Greenberg tells the story of Utah's Senator Orrin Hatch, a "standard, anti-Washington, budget-cutting conservative," finding a lump under his arm and calling the National Cancer Institute. They told him to "come right out there." It was diagnosed as a fatty deposit. Ever since, Hatch declared, he has been a big supporter of NIH, in tandem with liberal Democrat Henry Waxman of California.[1]

(Greenberg's book is "odd" because he first demonstrates the entanglement of science and politics and then criticizes scientists for not being political *enough*. Greenberg himself is a man of the Left.)

Scientists have followed in the teachers' footsteps. Public education declined in quality even as the amount of taxpayers' money spent on it sharply increased. Step by step, the teachers and their unions learned they could put their own welfare ahead of the students'. And get away with it. President Bush was played for a sucker by the education lobby when he called for "no child left behind." For years, the decline in public education was construed as just another indicator that not enough money had been spent.

Science is heading down the same path. A problem is discerned, or invented, the government steps in, and then the problem seems to grow more serious even as more attention is paid to it. That suits many of the scientists just fine.

Leaf through *Science* magazine and you will see that the maintenance of government spending on science is perhaps its leading preoccupation. Budgets are a major topic, scrutinized week after week. A few recent headlines: "Tight Budgets Force Lab Layoffs." "Bush Victory Leaves Scars—and Concerns About Funding." "A Dangerous Signal to Science."[2]

(There was great concern in this editorial because the EPA and the National Science Foundation "actually had their funding reduced from FY 2004 levels.") Dozens of such articles are published every year.

Still, bigger government is not a particularly "scientific" response to any crisis. Rarely are problems "solved" that way. But government spending does help some people, including the recipients of grants, and those who administer them.

Flattery can work wonders, especially with prospective "donors." NIH buildings are named after congressmen who control the purse strings. The Mark O. Hatfield Clinical Research Center was named after the long-time chairman of the Senate Appropriations Committee after he vowed to then NIH director Harold Varmus that he would protect the agency from budget cuts. "We may fail, but if we fail we're going to die with our boots on," Hatfield vowed.[3] The balanced budget amendment duly failed by one vote.

The John Edward Porter Neuroscience Research Center was named after the Illinois congressman who in 1995 became chairman of the House appropriations subcommittee for NIH, the starting point for medical-research appropriations. He led a delegation of scientists to meet with House Speaker Newt Gingrich to plead for favored treatment for the NIH budget, and when budget cutting loomed in 1996, he "telephoned ten university presidents and urged them to enlist the members of their boards of trustees in behalf of NIH."[4]

After Congressman Louis Stokes of Ohio retired in 1998, a $75 million building on the NIH campus was named the Louis Stokes Laboratory Building. One of the founders of the Congressional Black Caucus, Stokes was present at the dedication ceremony in 2001, and said that when Congressman Porter told him that the building would be named after him he was "absolutely surprised and stunned." Until that moment he had "absolutely no idea what it would be like having a building bearing my name," on the campus "of the greatest biomedical research institution in the world." Just think, he said, "from a little boy growing up in the projects in Cleveland to having a building named after you at the National Institutes of Health."

Stokes's "humanity" was praised. There was a jazz ensemble, and a pastor from Ebenezer Baptist Church blessed the event. But somehow the NIH news story failed to mention that before control of Congress shifted to Republican hands in 1995, Stokes had chaired the same appropriations subcommittee that Porter had taken over. Even so, the acting director of the NIH gave the game away when she said of Stokes: "His word was his bond—you could take it to the bank. And we did, many times."[5]

Younger members of the Black Caucus in attendance surely got the message—keep the money flowing and you too can have a building named after you. The National Center on Minority Health and Health Disparities will furnish them with plenty of rationales to keep on pouring the cash into NIH coffers. Absent from these political shenanigans has been even the slightest trace of doubt about the underlying equation: more money will give us better science.

Philanthropists of old could give away their own money and have university buildings named after them, and yet still be dismissed as robber barons. Today, a congressman can give away other people's money and be memorialized as a Hero of Science.

In 1999, a new building, still under construction, was named the Dale and Betty Bumpers Vaccine Research Center. This time President Clinton, Health and Human Services secretary Donna Shalala, and NIH director Harold Varmus were on hand for the celebrations. Talk of an AIDS vaccine filled the air—as it has, fruitlessly, for twenty-five years and counting. Again, Senator Dale Bumpers of Arkansas had served on the relevant Senate appropriations subcommittee. He "worked hard to increase funding for efforts to improve and purchase vaccines," according to the *Body*, in an article written for that publication by the NIH.

Bumpers's efforts "often resulted in Congress approving amounts above the Administration's requests."[6]

Congressmen are not usually memorialized until they retire, but an exception was made in the case of Senator Arlen Specter, whose name was prefixed to the National Library of Medicine in 1999. It, too, is on the NIH campus. Committee chairmen change quite frequently, so it is no surprise to learn that new buildings are going up all the time. The NIH building budget, a mere $190 million in 1998, had risen to $632 million by 2002. After twenty or thirty years, it's time for old buildings to be torn down anyway, so there will always be opportunities to keep on memorializing future allies in the budget wars.

Another influential figure who can reasonably expect a building sooner rather than later is House Speaker Dennis Hastert, who will not be outbid by Democrats in a race to expand the NIH. Hastert was "proud to announce" in 2001 that "the House Republican budget will double the funding for the National Institutes of Health by 2003 and will make history this year by providing the largest dollar increase in NIH's budget." There you have a Republican leader *boasting* about spending increases. Expect New Mexico's Pete Domenici, chairman of the Senate Energy Committee, to be memorialized on the 320-acre campus when he retires. He played a major role in funding the Human Genome Project (which started life in the Department of Energy). The project's budget grew from $28 million in 1988 to $500 million today.

Did I mention that most scientists are liberal Democrats? University-affiliated scientists are overwhelmingly so. The 1999 North American Academic Study Survey, including 1,643 U.S. faculty members, showed that in physics, liberals outnumber conservatives 6 to 1; in biology, 4.5 to 1. (Engineers are more evenly divided.)[7]

But when it comes to funding medical research, it doesn't make any difference what party you belong to.

Scientists are peddling hope as well as fear. There's a growing utopian inclination to believe that relief from the human condition—disease, aging, and perhaps even death itself—can be engineered with the latest technology. Spare body parts and replacement tissue may be created by bioengineers, much as mechanical engineers rebuild an automobile. Again, though, taxpayers are expected to foot the bill, which is where politics comes in. The stem cell hullabaloo boils down to the single issue of getting the federal government to pay for research that doesn't look too attractive to venture capitalists.

Scientists like to see themselves as motivated by idealism, but self-interest is not far behind. Their embrace of politics has undermined the objectivity that is supposed to be central to science. Day-to-day concerns about their own funding and security, and the fate of their latest grant proposals, overwhelm the more abstract concerns they may once have had about the integrity of the scientific method.

They have learned to "game the system," in other words. Scientists didn't start out that way, any more than teachers did. But slowly, year by year, they learned to consult their own advantage: discern a crisis, set up a hue and cry, send out press releases, reward friendly journalists with a heads-up about upcoming results that look newsworthy.

The media have cooperated with the scares, the hope, and the hype. Crisis sells newspapers. So the relationship is symbiotic. More important, journalists never for one minute doubt that more spending on any problem will reduce the magnitude of that problem. They are as devoted as any NIH official is to government spending as the cure for all ills. Consider the case of ABC News correspondent Sam Donaldson.

Advocates for increased spending on biomedical research discussed their strategy at an NIH event in 1998. Among the speakers was Donaldson, who had been treated for melanoma by the National Cancer Institute. "There is no shame in scientists' descent from the ivory tower to pitch a story to the press," he said. "Do not let your light be hidden under a basket. It is not beneath you to be originators, to come to say to us, 'Have I got a story for you!'"[8] Better to be "full of cash for medical research" than to be "poor and pristine."

I asked at the beginning of this book: where are Woodward and Bernstein? When it comes to science, they have joined the other side. They're team players with the government whose unchecked power once so concerned them. As for medical research, it is indeed "full of cash," and the underlying science is neither poor nor pristine. The only problem is that it has not been producing the promised results. And I hazard that it will not, as long as we so uncritically equate science funding and government funding. Governments can't do science. That was true in the Soviet Union, and it's true in the U.S., too.

NOTES

Introduction

The Lures of Politics

1. Barry Commoner, "Unraveling the DNA Myth: The Spurious Foundation of Genetic Engineering," *Harper's*, February, 2002.

Chapter 1

Global Warming

1. Willie Soon, Sallie Baliunas, et al., "Climactic and Environmental Changes of the Past 1,000 Years," *Energy and Environment*, vol. 14, 2003.

2. See David Deming, "Global Warming, the Politicization of Science and Michael Crichton's 'State of Fear'," in Fred Singer's *The Week That Was*, SEPP, March 5, 2005.

3. Antonio Regalado, "Global Warring," *Wall Street Journal*, February 14, 2005.

4. Quoted in Fred Singer's *The Week That Was*, February 19, 2005.

5. Robert Matthews, "Leading Scientific Journals Are 'Censoring Debate on Global Warming'," *Sunday Telegraph* (London), May 1, 2005.

6. Author interview with Ebell, April 2005.

7. Nicholas Kristof, "I Have a Nightmare," *New York Times*, March 12, 2005.

Chapter 2

Yes, More Nukes

1. President Eisenhower, "Atoms for Peace" speech, December 8, 1953.

2. Lewis L. Strauss, speech to National Association of Science Writers, New York, September 16, 1954.

3. Howard Hayden, e-mail to author, August 2005.

4. Edwin Newman, Earth Day, 1970.

5. Amity Shlaes column, *Financial Times* (London), April 28, 2005.

6. "Experts Find Reduced Effects of Chernobyl," *New York Times*, September 6, 2005.

7. David Albright et al., *Plutonium and Highly Enriched Uranium 1996: World Inventories, Capabilities and Policies* (Oxford: Oxford University Press, 1997).

8. "China Promotes Another Nuclear Boom," *New York Times*, January 15, 2005.

9. Petr Beckmann, *The Health Hazards of Not Going Nuclear* (Boulder, CO: Golem Press, 1977.)

10. Peter Huber and Mark Mills, "Why the U.S. Needs Nuclear Power, *City Journal*, Winter 2005.

11. Spencer Abraham, talk at Competitive Enterprise Institute, *American Enterprise*, September 2001.

12. Howard Hayden, *The Solar Fraud* (Pueblo West, CO: Vales Lake Publishing, 2004.)

13. "Are These Towers Safe?" *Time*, June 20, 2005.

14. See Center for Biological Diversity, Endangered Earth Online, #360: "Judge OKs Lawsuit Against Killer Wind Turbines," March 2, 2005.

15. Howard Hayden, e-mail to author, June 2005.

16. *Nevada Review Journal*, August 11, 2005.

17. "Cronkite Withdraws Ad Against Turbines," *Vineyard Gazette*, August 28, 2003; see also http://www.CapeWind.org.

18. "Researchers Alarmed by Bat Deaths from Wind Turbines," *Washington Post*, January 1, 2005.

19. See Hayden, *Solar Fraud*.

20. Bill McKibben, "Tilting at Windmills," *New York Times*, February 16, 2005.

21. Stewart Brand, "Environmental Heresies," *Technology Review*, May 2005; see also "Old Foes Soften to New Reactors," *New York Times*, May 15, 2005.

Chapter 3

Good Vibes: The Virtue of Radiation

1. For his resume, log on to http://www.ratical.org/radiation/CNR/JWGcv.html.

2. "With Radiation, How Little Is Too Much?" *New York Times* (Week in Review), September 26, 1982.

3. "For Radiation, How Much Is Too Much?" *New York Times*, November 27, 2001.

4. Joby Warrick, *Washington Post*, April 14, 1997.

5. Edward B. Lewis, *Science*, v. 125, May 17, 1957.

6. Author interview with Theodore Rockwell, Bethesda, Maryland, 2002.

7. See "Nuclear Shipyard Worker Study (1980–1988): A large cohort exposed to low dose-rate gamma radiation." It is available on the "Radiation, Science and Health" website.

8. "Does Radiation Exposure Produce a Protective Effect Among Radiologists?" *Health Physics*, v. 52, 1987, 637–43; "Radiation Increased the Longevity of British Radiologists," *British Journal of Radiology*, v. 75, 2002, 637–38; see also SEPP website, October 16, 2004, "Longevity and Radiation," by John Cameron.

9. "Ukraine Consents to Shut Chernobyl Before Year's End," *New York Times*, June 6, 2000.

10. *New York Times*, September 6, 2005.

11. Bernard L. Cohen, "The Myth of Plutonium Toxicity," *Health Physics*, v. 32, 1977, 359–79, discussed by Cohen in Karl Otto Ott and Bernard I. Spinard, eds., *Nuclear Energy* (New York: Plenum Press, 1985), 355–65.

12. Ibid.

13. Bernard L. Cohen, *Health Physics*, v. 68, 1995, 359–79; and in J. Radiol. Prot. V. 19, 1999, 63–65.

14. Author interview with Cohen, 2002.

15. See W. L. Chen, Y. C. Luan, et al., "The Immune Effects of Radiation Observed from the Incident of Co-60 Contaminated Apartments in Taiwan," BelleOnline, international conference, "Non-linear Dose-response Relationships in Biology, Toxicology and Medicine," May 28–30, 2003. See BELLE website, http://www.belleonline.com.

16. See "Very High Background Radiation Areas of Ramsar, Iran: Preliminary Biological Studies," *Health Physics*, v. 82, 2002, 87–93; see also BELLE website.

17. Author interview with Klaus Becker, 2002.

Chapter 4

"Good Chemistry"

1. Quotes taken from author's frequent interviews with Ed Calabrese, 2002, 2003.

2. E. J. Calabrese and L. Baldwin, "Toxicology Rethinks Its Central Belief," *Nature*, v. 421, February 13, 2003.

3. Leonard Sagan and Sheldon Wolff, "On Radiation, Paradigms and Hormesis," *Science*, August 11, 1989, 574, 575.

4. Will Hively, "Is Radiation Good for You? Or Dioxin? Or Arsenic?" *Discover*, December 2002.

5. For much background information on hormesis, see Edward Calabrese in *Human and Experimental Toxicology*, v. 19, January 2000, "Special Issue on Hormesis."

6. See Michael Gough, *Dioxin, Agent Orange: The Facts* (New York: Plenum Press, 1986), 121.

7. See Michael Fumento, *Science Under Siege* (New York: William Morrow & Co.), 1996.

8. Ibid., 111.

9. Ralph Cook, "Responses in Humans to Low Level Exposures," in *Biological Effects of Low Level Exposures: Dose Response Relationships* (New York: Lewis Publishers, 1994), 102.

10. Fumento, 112.

11. Ibid.

12. Cook, 107; *New York Times*, August 15, 1991, "Times Beach Warning: Regrets a Decade Later."

Chapter 5

The DDT Ban

1. "Fighting Malaria with DDT," *New York Times*, December 23, 2002; "Death by Environmentalist," *Wall Street Journal*, December 29, 2004.

2. See Steven Milloy, *Junk Science Judo*, Cato Institute, 2001, and Malloy's material on DDT available at http://www.junkscience.com.

3. Rachel Carson, *Silent Spring* (Boston: Houghton Mifflin Co., 1962), reprinted in 2000 with an afterword by Edward O. Wilson; and see Lisa Makson, "Rachel Carson's Ecological Genocide," FrontPageMagazine.com, July 31, 2003.

4. See J. Gordon Edwards and Steven Milloy, "100 Things You Should Know About DDT," http://www.junkscience.com/ddtfaq.

5. J. Gordon Edwards, "DDT: A Case Study in Scientific Fraud," *Journal of American Physicians and Surgeons*, v. 9, no 3, fall 2004, 83–88.

6. Roy Innis letter to President Bush, November 2004.

7. Makson, "Rachel Carson's Ecological Genocide."

8. See Michael McCloskey in J. Gordon Edwards, "Malaria, the Killer that Could Have Been Conquered," *21st Century Science and Technology*, summer 1993.

9. "*New York Times* Favors DDT," A.I.M. *Media Monitor*, January 6, 2003.

10. "Wolfowitz Discusses Issues with Bono," Reuters, March 18, 2005.

Chapter 6

Biodiversity and Endangered Species

1. E. O. Wilson, *The Future of Life* (New York: Knopf, 2003), 98–99.

2. Ibid., xxii.

3. Ibid., 131.

4. "The Sixth Extinction," *National Geographic*, February 1999.

5. See Patrick Moore, "Environmentalism for the 21st Century," http://www.greenspirit.com, 16–18.

6. Bjørn Lomborg, *The Skeptical Environmentalist* (Cambridge: Cambridge University Press, 2001).

7. Norman Myers, *The Sinking Ark* (Oxford: Pergamon Press, 1979).

8. Matt Ridley, "The Profits of Doom," *American Spectator*, February 23, 2002.

9. See *Scientific American*, January 2002, and Lomborg's reply, May 2002.

10. David Quammen, "Planet of the Weeds," *Harper's*, October 1998.

11. Moore, "Trees Are the Answer," http://www.greenspirit.com, 5.

12. Lawrence Slobodkin, "Islands of Peril and Pleasure," *Nature*, May 16, 1996, 205.

13. Rowan B. Martin, "Biological Diversity," in Ronald Bailey, ed., *Earth Report 2000* (Chicago: McGraw-Hill, 2000).

14. Charles Mann and Mark Plummer, *Noah's Choice: The Future of Endangered Species* (New York: Knopf, 1995), 75–76.

15. Michael Sanera and Jane S. Shaw, *Facts Not Fear* (Washington, D.C.: Regnery, 1996), 118–19; *Wall Street Journal*, August 3, 2005.

16. "Two New Primate Species Discovered," *National Geographic*, June 24, 2002.

17. "New Monkey Species Discovered in Africa," *New Scientist*, May 19, 2005.

18. Wilson, 91.

19. "House Votes for New Limits on Endangered Species Act," *New York Times*, September 30, 2005.

Chapter 7

African AIDS: A Political Epidemic

1. WHO, "Workshop on AIDS in Central Africa," October 22–25, 1985.

2. WHO, *Weekly Epidemiological Record*, No. 10, March 7, 1986.

3. *Science*, November 21, 1986.

4. Mark Schoofs, "AIDS: The Agony of Africa," *Village Voice*, eight parts, November 1999–January 2000.

5. *New York Times*, July 3, 1981.

6. Lawrence K. Altman, "AIDS in Africa: A Pattern of Mystery," *New York Times*, November 8, 1985; see also *New York Times*, November 21, 1985, and December 22, 1985.

7. Joseph McCormick and Susan Fisher-Hoch, *Level 4: Virus Hunters of the CDC* (New York: Barnes & Noble, 1999), 189.

8. Ibid., 190.

9. Ibid., 176.

10. WHO, *Weekly Epidemiological Record*, November 26, 1999.

11. Rian Malan, "AIDS in Africa: In Search of the Truth," *Rolling Stone*, November 22, 2001.

12. South Africa Department of Health, Summary Report: *1998 National HIV Sero-prevalence Survey of Women Attending Public Antenatal Clinics in South Africa*, February 1999. Cited in Charles Geshekter, Sam Mhlongo, and Claus Kohnlein, "AIDS, Medicine and Public Health," presented at the 47th annual meeting of the African Studies Association, November 11, 2004.

13. Charles Gilks, "What Use Is a Clinical Case Definition for AIDS in Africa?" *British Medical Journal*, 303 (1991), 1190.

14. "Dead Zones" series, eight parts, *New York Times*, August 6, 1998, to December 27, 1998.

15. *New York Times*, December 16, 1998.

16. Erik Eckholm and John Tierney, "AIDS in Africa: A Killer Rages On," *New York Times*, September 16, 1990. First of a four-part series.

17. Andy Rooney, "How Much Is Enough?" Tribune Media Services, July 14, 2005.

18. Rian Malan, "Pugwash Hogwash," *Spectator*, October 2, 2004.

19. Population Reference Bureau, "World Population Highlights, 2004."

20. *New York Times*, August 6, 2005.

21. *Daily Telegraph* (London), July 6, 2005.

22. *Daily Mail* (London), July 20, 2005.

Chapter 8

The Folly of Dolly: Cloning and Its Discontents

1. Gina Kolata, *Clone: The Road to Dolly and the Path Ahead* (New York: Allen Lane/The Penguin Press, 1997), 100–01.

2. Ibid., 119.

3. "Some Scientists Ask: How Do We Know Dolly Is a Clone?" *New York Times*, July 29, 1997.

4. "Creator of Cloned Sheep Says He Will Try to Repeat Process," *New York Times*, February 14, 1998.

5. "Beating Hurdles, Scientists Clone a Dog for a First," *New York Times*, August 4, 2005.

6. "In a Furry First, A Dog Is Cloned in South Korea," *Washington Post*, August 4, 2005.

7. *New York Times*, May 27, 1999.

8. *New York Times*, March 25, 2001.

9. *New York Times*, February 15, 2002.

10. *Washington Post*, August 4, 2005.

11. "In Cloning, Failure Far Exceeds Success," *New York Times*, December 11, 2001.

12. Stephen S. Hall, "The Recycled Generation," *New York Times Magazine*, January 30, 2000.

13. Ibid.

Chapter 9

The Stem Cell Challenge to Bioengineering

1. Associated Press, February 16, 2001.

2. BBC News, February 17, 2001.

3. Parkinson's Information Exchange website.

4. *Science*, v. 291, March 2, 2001.

5. *Science*, July 26, 2002.

6. Gina Kolata, *New York Times*, March 8, 2001.

7. Gina Kolata, "A Cautious Awe Greets Drug That Eradicates Tumors in Mice," *New York Times*, May 3, 1998.

8. *Washington Post*, November 6, 1998.

9. Connie Bruck, "Hollywood Science," *New Yorker*, October 18, 2004.

10. Nicholas Wade, "Tracking the Uncertain Science of Growing Heart Cells," *New York Times*, March 14, 2005.

11. Robert Lanza and Nadia Rosenthal, "The Stem Cell Challenge," *Scientific American*, June 2004.

12. *Washington Post*, November 6, 1998.

13. *Science*, December 17, 1999.

14. *New York Times*, April 27, 2001.

15. *Wall Street Journal*, August 12, 2004.

16. "Stem Cells Not So Stealthy After All," *Science*, v. 297, July 12, 2002.

17. Stephen Jay Gould, "What Only the Embryo Knows," *New York Times*, August 27, 2001.

Chapter 10

A Map to Nowhere

1. W. French Anderson, "Gene Therapy," *Scientific American*, September 1995.

2. "DNA Mapping Milestone Heralded," *Washington Post*, June 27, 2000.

3. Venter opening statement at Washington press conference, February 12, 2001.

4. See "Not Junk After All," *Science*, May 23, 2003; "Spinning Junk Into Gold," *Science*, June 13, 2003.

5 *New York Times*, June 25, 2000.

6 *Nature*, February 15, 2001.

7. "Biotech CEO Says Map Missed Much of Genome," *Boston Globe*, April 9, 2001.

8. Andrew Pollack, "Double Helix with a Twist; Do Fewer Genes Translate Into Fewer Dollars?" *New York Times*, February 13, 2001.

9. Eric Lander on *News Hour*, PBS, April 25, 2003.

10. "Cystic Fibrosis Gave Up Its Gene 12 Years Ago," *Wall Street Journal*, June 11, 2001.

11. "Gene Therapy Apparently Cures 2," *Washington Post*, April 28, 2000.

12. "Scientists Report the First Success of Gene Therapy," *New York Times*, April 28, 2000; and editorial.

13. Rick Weiss, "Dream Unmet After 50 Years," *Washington Post*, February 28, 2003.

14. Sharon Begley, "Science Journal," *Wall Street Journal*, May 23, 2003.

15. James Shreeve, *The Genome War* (New York: Knopf, 2004), 362.

16. Ibid., 360.

17. James Shreeve, "Craig Venter's Epic Voyage to Redefine the Origin of Species," *Wired*, August 2004.

Chapter 11

The Great Cancer Error

1. Dorothy Nelkin and M. Susan Lindee, *The DNA Mystique: The Gene As Cultural Icon* (New York: W. H. Freeman, 1995).

2. Andrew Pollack, "Huge Genome Project Is Proposed to Fight Cancer," *New York Times*, March 28, 2005.

3. Clifton Leaf, "Why We're Losing the War on Cancer (And How to Win It)," *Fortune*, March 22, 2004.

4. Robert A. Weinberg, *Racing to the Beginning of the Road* (New York: W. H. Freeman, 1998), 121.

5. See Peyton Rous, Nobel Lecture, December 13, 1966, at http://www.nobelprize.org.

6. P. H. Duesberg, "Retroviruses as carcinogens and pathogens: expectations and reality," *Cancer Research*, v. 47, 1987. Reprinted in P. H. Duesberg, *Infectious AIDS: Have We Been Misled?* (Berkeley: North Atlantic Books,1995).

7. J. Michael Bishop, *How to Win the Nobel Prize* (Boston: Harvard University Press, 2003), 162.

8. *New York Times*, October 10, 1989.

9. Jean Marx, *Science*, October 20, 1989.

10. *Washington Post*, December 11, 2004.

11. *New York Times*, March 28, 2005.

12. Robert A. Weinberg, *One Renegade Cell* (New York: Basic Books, 1998), 50–51.

13. Theodor Boveri, *The Origin of Malignant Tumors*, 1914. English translation, Baltimore, MD: Williams and Wilkins, 1929.

14. W. Wayt Gibbs, "Untangling the Roots of Cancer," *Scientific American*, July 2003.

15. Author interview with Christoph Lengauer, 2001.

Chapter 12

The Abiding Myths

1. Jeffrey Burton Russell, *Inventing the Flat Earth* (Westport, CT: Praeger, 1991), 4.

2. Stephen Jay Gould, *Dinosaur in a Haystack* (New York: Harmony Books, 1995), 95.

3. J. B. Russell, 26.

4. C. S. Lewis, *The Discarded Image* (Cambridge: Cambridge University Press, 1964), 140–41.

5. J.B. Russell, 53.

6. Ibid., 28.

7. David C. Lindberg and Ronald L. Numbers, eds. *God and Nature: Historical Essays on the Encounter Between Science and Christianity* (Berkeley, CA: University of California Press, 1986), 2.

8. James R. Moore, *The Post-Darwinian Controversies* (Cambridge: Cambridge University Press, 1979), 36.

9. J. B. Russell, 42.

10. Lindberg and Numbers, 3.

11. Bertrand Russell, *Religion and Science* (1935) (New York: Oxford University Press, 1961), 7.

12. Ibid., 24.

13. John Hedley Brooke, *Science and Religion—Some Historical Perspectives* (Cambridge: Cambridge University Press, 1991), 85–88.

14. H. W. Crocker III, *Triumph* (New York: Prima Publishing, 2001), 310.

Chapter 13

By Chance, or by Design?

1. Bush statement from the White House, August 1, 2005.

2. Christoph Schonborn, "Finding Design in Nature," *New York Times*, July 7, 2005.

3. Richard Dawkins, *The Blind Watchmaker* (New York: W. W. Norton, 1996), 1, 6.

4. Daniel Dennett, *Darwin's Dangerous Idea* (New York: Simon & Schuster, 1995), 63.

5. Michael Powell, "Doubting Rationalist," *Washington Post*, May 15, 2005.

6. Cornelia Dean, "Opting Out in the Debate on Evolution," *New York Times*, June 21, 2005.

7. "Evolution Wars," *Time*, August 15, 2005.

8. Kenneth Miller, *Finding Darwin's God* (New York: HarperCollins, 2000), 53–54.

9. Kenneth Chang, "In Explaining Life's Complexity, Darwinists and Doubters Clash," *New York Times*, August 22, 2005.

10. Donald Kennedy, "Twilight for the Enlightenment?" *Science*, April 8, 2005.

11. Julian Huxley, in Sol Tax, ed., *Evolution After Darwin*, vol. 3 (Chicago: University of Chicago Press, 1960), 111.

12. Garrett Hardin, *Nature and Man's Fate* (Dublin: Mentor Books, 1961), 216.

13. Thomas Hunt Morgan, *Evolution and Genetics* (Princeton, NJ: Princeton University Press, 1925), 120, 127.

14. C. H. Waddington, "Evolutionary Adaptation," in Sol Tax, ed., *Evolution After Darwin*, vol. 1 (Chicago: University of Chicago Press, 1960), 385.

15. Karl Popper, *Unended Quest: An Intellectual Biography* (Oxford: Routledge, 1992), 199.

16. William Dembski, *Intelligent Design* (Downers Grove, IL: InterVarsity Press, 1999), 229.

17. Bertrand Russell, *Religion and Science* (Oxford: Oxford University Press, 1935), 72–73.

18. Richard Lewontin, *Biology as Ideology* (New York: HarperPerennial, 1991), 10.

19. Richard Dawkins, *A Devil's Chaplain* (London: Orion Books, 2004), 94.

20. Quoted in Michael Behe, *Darwin's Black Box* (New York: The Free Press, 1996); Waddington, "Evolutionary Adaptation," in Tax, ed., 385; Popper, 199.

Chapter 14
Evolution: The Missing Evidence

1. Letter from Colin Patterson to Luther Sunderland, April 10, 1979. The edited transcript of Patterson's talk at the Natural History Museum is available online at the Access Research Network website, http://www.arn.org. Many of those who asked questions from the audience are identified. The second edition of Patterson's *Evolution* was published by Cornell University Press in 1999.

2. Ernst Mayr, *The Growth of Biological Thought* (Cambridge, MA: Harvard University Press, 1982), 232, 465.

3. Quotations from Jonathan Wells are taken either from his book *Icons of Evolution: Science or Myth* (Washington, D.C.: Regnery, 2000), or from his article "The Survival of the Fakest," *American Spectator*, December 2000/January 2001.

4. Nancy B. Simmons, "An Eocene Big Bang for Bats," *Science*, January 28, 2005.

5. Henry Gee, *In Search of Deep Time* (New York: The Free Press, 1999).

6. Tim Berra, *Evolution and the Myth of Creationism* (Stanford: Stanford University Press, 1990), 117.

7. Wells, *Icons of Evolution*, 68.

8. Phillip Johnson, *Defeating Darwinism by Opening Minds* (Downers Grove, IL: InterVarsity Press, 1997), 63.

9. Elizabeth Pennisi, "Haeckel's Embryos: Fraud Rediscovered," *Science*, September 5, 1997.

10. Stephen Jay Gould, "Abscheulich! Atrocious!" *Natural History*, March 2000.

11. H. B. Kettlewell, "Darwin's Missing Evidence," *Scientific American*, March 1959; see also Wells, *Icons of Evolution*, chapter seven.

12. W. Ford Doolittle, "Phylogenetic Classification and the Universal Tree," *Science*, v. 284, 1999, 2124–128.

13. See Wells, *Icons of Evolution*, chapter two.

14. Phillip Johnson, "The Church of Darwin," *Wall Street Journal*, August 16, 1999; Wells, chapter eight.

15. Charles Darwin, *On the Origin of Species* (1859) (New York: Penguin Books, 1968), 133.

16. See also Emmett Reid Dunn, "Evolution While You Watch," *American Mercury*, May 1927.

17. Quoted by Norman Macbeth in *Darwin Retried* (New York: Delacorte Press, 1973), 36.

18. Jonathan Wells, e-mail to author, May 2005.

Final Thoughts

1. Daniel S. Greenberg, *Science, Money, and Politics* (Chicago: University of Chicago Press, 2001), 439.

2. *Science*, July 15, 2005; November 12, 2004; December 24, 2004.

3. "Scientists Mobilize to Fight Cuts," *Science*, May 26, 1995.

4. Greenberg, 439.

5. "Stokes Honored by Lab Building Dedication," *NIH Record*, July 10, 2001.

6. "NIH Dedicates New Vaccine Research Center to Dale and Betty Bumpers," *Body*, June 9, 1999.

7. Stanley Rothman, S. Robert Lichter, Neil Nevitte, "Politics and Professional Advancement Among College Faculty," *Forum*, vol. 3, no. 1, 2005.

8. American Society of Cell Biology Newsletter, June 1998.

INDEX

Abbott Laboratories, 114, 115

Abelson, Philip, 77

Abraham, Spencer, 26

acquired immune deficiency syndrome (AIDS): in Africa, ix, 105–21; condom distribution programs and, 117–18; consensus and, 112; deaths from, 115–18; defining, 107–13; diagnosis of, 113–14; Hollywood and, 113; media and, 105, 110; politicization of science and, 105; population and, 118–19; symptoms of, 107–8; testing for, 113–14; in United States, 106, 108

Advanced Cell Technology (ACT), 128–29, 143

Africa: AIDS in, 105–21; DDT in, 73; HIV/AIDS in, ix; malaria in, 73–74, 79–81; population of, ix, 100, 105; post-colonial health in, 119–21

Agent Orange, 64, 66

"AIDS in Africa, in Search of the Truth" (Malan), 106, 117

AIDS: The Failure of Contemporary Science (Hodgkinson), 119

Alberts, Bruce, 229

Albright, Madeleine, ix

Altamont Pass, Calif., 27, 30

Altman, Lawrence K., 110–11, 237

Alzheimer's disease, stem cell research and, 134, 136, 140

American Association for the Advancement of Science, 131

American Journal of Epidemiology, 70

Ames, Bruce, 51, 76, 168

Anderson, W. French, 148

antibiotics, 60–61

Appleyard, Bryan, 129

Aquinas, Thomas, 183

Arndt, Rudolph, 61, 62

Arndt-Schulz Law, 61

Asia, malaria in, 74

The Assayer (Galileo), 194

Atomic Bomb Casualty Commission, 44

Atomic Energy Commission, 20, 43

Audubon Society, 78

Australia, Kyoto Protocol and, 4

Austria, radiation and, 55

Bacon, Roger, 183

Baer, Karl Ernst von, 145

Baliunas, Sally, 8

Baltimore, David, 152, 169

Barber, Lionel, 135

Bat Conservation International, 33

Beadle, George, 151

Becker, Klaus, 54–55

Beckmann, Petr, 25

Begley, Sharon, 161

Behe, Michael, 199, 202, 212–14, 225
Bellarmine, Cardinal, 194, 196
Berlinski, David, 210–11, 213, 214
Bernstein, Carl, vii, 243
Berra, Tim, 225
Bill and Melinda Gates Foundation, 84
biodiversity, 162; extinction and,
    87–89, 96; global warming and, 4;
    overpopulation and, 88
bioengineering: cloning and, 123–30;
    stem cell research and, 131–45
Biological Effects of Low Level Expo-
    sure (BELLE), 59
Bishop, J. Michael, 169, 170–71
Bitman, J., 77–78
The Blind Watchmaker (Dawkins), 199
Bono, 84, 119, 121
Boorstin, Daniel, 182
Boston Globe, 155
Boveri, Theodor, 176
The Boys from Brazil (Levin), 123
Bradley, James, 196
Brahe, Tycho, 195–96
Brand, Stewart, 36
Brandt, Edward, 112
Brave New Worlds: Staying Human in
    the Genetic Future (Appleyard), 129
Bray, Dennis, 15
British Medical Journal, 115
Brooke, John Hedley, 191, 195, 196
Browner, Carol, 34
Bruno, Giordano, 196
Bumpers, Dale, 240–41
Burbank, Luther, 233–34
Buridan, Jean, 183
Bush, George W., vi, 80, 84, 133–34,
    199, 238
Butler, Philip, 77
Byers, Eben M., 63

Calabrese, Edward, hormesis and,
    58–63
Calgary Herald, 5

California–Berkeley, University of,
    40–41, 76, 77, 96
California Proposition 71, 128, 137,
    139, 143
California–San Francisco, University
    of, 135
Caltech, 13, 43, 152
Calvin, John, 192
cancer: aneuploidy and, 165, 176–80;
    cancer genome project and, x,
    165–66, 173; DDT and, 75–76, 78;
    dioxin and, 67–71; gene-mutation
    theory of, xii, xiii, 163, 167–80;
    media and, 166–67; nuclear power
    and, 32; origins of, xii, 165–80; politi-
    cization of science and, 180; radia-
    tion and, 39, 42, 43–44, 48, 51, 52,
    60, 63, 168; stem cell research and,
    131, 135; viruses and, 168–70, 175
Cape Wind Project, 31–33
Carson, Rachel, 74–75, 79
Carter, Jimmy, 3–4, 26, 33
Catholic Church: flat earth myth and,
    185–86; intelligent design and, 199;
    science and religion and, 181,
    189–92, 195–97
CDC. See U.S. Centers for Disease Con-
    trol (CDC)
CEI. See Competitive Enterprise Insti-
    tute
Celera Genomics, 148, 162
Center for Biological Diversity (CBD),
    30
Center for Ecology and Hydrology
    (Britain), 50
Center for Low Dose Research, 55
Central African Republic, 105, 107
Chernobyl, Ukraine, 21, 28, 47, 48–50
"Chernobyl's Legacy" (UN), 48, 49
China: DDT and, 82; Kyoto Protocol
    and, 4; nuclear power and, 21, 25
The China Syndrome, 21
chloracne, 66

cladistics, evolutionary theory and, 218–19

clean air: Clean Air Act and, 3; politicization of science and, viii

Clean Air Act, 3

Clean Water Act, 3

Clinton, Bill, 4, 26, 34, 148, 169

Clone: The Road to Dolly and the Path Ahead (Kolata), 124

cloning, 123–30; problems with, 123, 126–28; stem cell research and, 129–30

Cohen, Bernard, 25, 50–53

Collins, Francis S., 148, 149, 151, 152, 155

Colorado, 19, 20, 51

Columbus, Christopher, 183–84, 185, 188

Commentary, 213

Commoner, Barry, 44

Competitive Enterprise Institute (CEI), 16, 100

Condillac, Etienne Bonnot de, 183

Congressional Black Caucus, 80, 239, 240

Congress of Racial Equality, 80

consensus: AIDS and, 112; evolution and, xiii; science and, xiii, 13–14

Cook, Ralph, 69, 71

Cookson, Clive, 135

Cooper Institute, 186

Copernicus, 190–93, 195–96

Coyne, Jerry, 203

Crichton, Michael, vii, viii, 13–14, 16

Crick, Francis, 149, 154, 156, 161, 212

Cronkite, Walter, 31–32

cystic fibrosis, 147, 157, 160

Cystic Fibrosis Foundation, 158

Daily Mail, 120

Daily Telegraph (London), 120

Darwin, Charles, 143, 205, 208, 215, 227; extinction and, 89; homology and, 220; science and religion and, 187–89. *See also* evolutionary theory

Darwinism. *See* Darwin, Charles; evolutionary theory

Darwin's Black Box (Behe), 202, 212, 214

Darwin's Dangerous Idea (Darwin), 200

Davis, Patti, 134

Dawkins, Richard, 199–200, 201, 209, 211, 213

DDT, 73–85; banning of, 73, 78–79; birth control and, 76; bringing back, 73, 84–85; cancer and, 75–76, 78; eggshells and, 77–78; EPA and, 73, 75–76, 78–79; malaria and, 73–74, 79–81, 83; media and, 73; politicization of science and, 75–77, 82–83; public policy and, 83

"The Death of Environmentalism" (Shellenberger and Nordhaus), 17

de Beer, Sir Gavin, 207

Dembski, William, 208, 214, 232

Deming, David, 8–9

Dennett, Daniel, 200, 201, 208, 209, 210–11, 213

Descartes, Rene, 197

The Design Inference: Eliminating Chance Through Small Probabilities (Dembski), 232

determinism, x; genetic, 88

diabetes, stem cell research and, 136, 140–42

Dialogue on the Two Chief Systems of the World (Galileo), 194

Diderot, Denis, 183

dioxin, 57; cancer and, 67–71; EPA and, 64, 65, 67, 69; hormesis and, 65; panic over, 63–65; studying, 65–71

Dioxin, Agent Orange: The Facts (Gough), 70

The Discipline of Curiosity (King), 76

The Discoverers (Boorstin), 182

Discovery Institute, 202

Dolly, 124, 125–27

Domenici, Pete, 149, 241
Dominko, Tanja, 128
Donaldson, Sam, 242–43
Doolittle, W. Ford, 228
Dow Chemical Company, 64, 67
Down's Syndrome, 174
Draper, John William, 184–86, 187, 187–88
Duesberg, Peter H., 119, 169–70, 173, 174–75, 177–80
DuPort, Philippe, 55
Duquesne Light, 19

Earth Day (1970), 3
Earth Summit (1992), 4
Ebell, Myron, 16
ecology, 74
economics, vii; global warming and, 1; science and, viii
Edwards, J. Gordon, 77, 78
Ehrlich, Paul, ix, 83, 93, 101
Einstein, Albert, 191
Eisenhower, Dwight D., 19, 20, 44
Eliasson, Jan, 120
Ellison, Michael, 150
endangered species: Endangered Species Act and, 3, 93; political taxonomy and, 99–100; politicization of science and, viii; property issues and, 100–103. *See also* extinction
Endangered Species Act, 3, 93
"The Endangered Species Act: A Perverse Way to Protect Biodiversity" (Stroup), 93
Endangered Species Act of 1973, 99, 102, 103
energy: hydroelectricity, 34; nuclear power, 19–37; renewable, 27, 29–30, 33–37; solar, 33
Energy Department, 47, 59
Eniwetok Atoll, 43
environmentalism, environmentalists: clean air and, viii, 3; endangered species and, viii, 3, 93, 99–103; extinction and, 89–92; global warming and, viii, 1–17, 26, 36–37; lobbying and, 16; politics and, 89; radiation and, 39; renewable energy and, 27, 29–30, 34–37
Environmental Protection Agency (EPA), x, 239; DDT and, 73, 75–76, 78–79; dioxin and, 64, 65, 67, 69; global warming and, 3; hormesis and, 57; nuclear power and, 32; radiation and, 39, 46, 51–52
Environment Canada, 11
Erwin, Douglas H., 202
Essay on the Principle of Population (Malthus), 209
eugenics, 13, 123
European Union (EU), 10
Evans, Martin, 134–35
evolutionary psychology, 88
evolutionary theory: bats and, 218–26; building blocks and, 228–29; cladistics and, 218–19; consensus and, xiii; evidence for, 215–35; extinction and, 89; Galapagos finches and, 229–31; Haeckel's embryos and, 226; homology and, 220–26; intelligent design vs., 199–205, 220; irreducible complexity and, 212–14; natural selection and, 199, 205–6, 207–9, 232; peppered moths and, 227; religion and, 187–89, 216; speciation and, 234–35; survival of the fittest and, 206; tree of life and, 227–28; weakness of, 147, 199–200, 205–12.
"The Evolution of Life On Earth" (Gould), 193
extinction: biodiversity and, 87–89, 96; cases of, 97–99; environmentalism and, 89–92; evolution and, 89; politicization of science and, 92–97; truth about, 89–92. *See also* endangered species

Facts not Fear: A Parent's Guide to Teaching Children about the Environment (Sanera and Shaw), 97

Farber, Celia, 105

Financial Times, 135

Finding Darwin's God (Miller), 201–2

Fishbein, Jonathan, vii

flat earth myth, 182–86

Foldman, Judah, 134

Fonda, Jane, 21, 22

Food and Drug Administration (FDA), 58, 161

Fortune, 60, 166

fossil fuels, 2, 10, 36

Franklin, Benjamin, 183

Freud, Sigmund, 193

Friends of the Earth, 36

From Darwin to Hitler, Evolutionary Ethics, Eugenics and Racism in Germany (Weikhart), 214

Fumento, Michael, 64

The Future of Life (Wilson), 87, 99

Galileo Galilei, 181, 191–92, 193–95, 196

gamma rays, 54–55

Garrett, Laurie, 118

Gates, Bill, 143

Gearhart, John D., 138

Gee, Henry, 223

Geiger counter, 24

Geldof, Bob, 119

Gelsinger, Jesse, 160

gene-mutation theory. *See* cancer

General Electric, 57

genetic engineering: gene therapy and, 148–49, 159, 159–61; genome decoding and, 149–53, 153–55, 161–63; heritable diseases and, 151–52, 157–59; Human Genome Project and, 147–48, 157–59; mutations and, 157–59; recombinant DNA and, 147–48; spending on, 147; worthless genes and, 155–56

The Genome War: How Craig Venter Tried to Capture the Code of Life and Save the World (Shreeve), 162

Germany, 11, 15, 66, 145; Kyoto Protocol and, 4; radiation and, 55

Geron Corporation, 128, 138, 139

Gibbon, Edward, 183

Gibbs, W. Wayt, 178–79

Gilbert, Walter, 125, 155

Gilks, Charles, 115

Gingrich, Newt, 239

global warming: biodiversity and, 4; causes of, 1–3; challenging, 8–17; earth surface temperature data and, 1, 2, 7–9; economics and, 1; EPA and, 3; greenhouse emissions and, 1–3, 26; "hockey stick" line and, 6–8, 9, 12; Kyoto Protocol and, 1, 4–5, 10, 26; media and, 9, 11; nuclear power and, 19, 36–37; politicization of science and, viii

Global Warming's Unfinished Debate (Singer), 15

"Global 2000" report, 3–4, 93

Gofman, John, 40–41, 45

Goldstein, Robert, 140, 141

Gore, Al, ix, 118

Gough, Michael, xiii, 70

Gould, Stephen Jay: Copernicus and, 191; evolutionary theory and, 188, 203, 209, 226; flat earth myth and, 182, 183; man's self-importance and, 193; stem cell research and, 145

Grant, Edward, 185

Grant, Peter and Rosemary, 229–30

Greenberg, Daniel S., 237–38

Greenhouse Effect, 2

greenhouse emissions: global warming and, 1–3, 26; Kyoto Protocol and, 1, 4–5

Green House Network, 13

Greenland, 3, 8

Green Party, 3, 26

Greenpeace, 36, 82, 90, 91, 93
Gregory, Dick, 22

Haeckel, Ernst, 154, 226
Haldane, J. B. S., 123
Hardin, Garrett, 101, 207
Harper's, x, 94
Harvard University, 8, 14, 42, 141
Haseltine, William, 155
Hastert, Dennis, 241
Hayden, Howard, 20, 25, 26, 29, 31, 34, 35
Hayden, Tom, 21
Hayflick, Leonard, 174
The Health Hazards of Not Going Nuclear (Beckmann), 25
Health Physics, 45, 53
Hennig, Willi, 219
Heubner, Robert, 169
Hind, Rick, 82
Hiroshima, Japan, 28, 39, 41, 43
A History of Civilization: Prehistory to 1715, 188
History of Conflict between Religion and Science (Draper), 184
History of the Life and Voyages of Christopher Columbus (Irving), 184–85
A History of the Warfare of Science with Theology in Christendom (White), 185
HIV. *See* human immunodeficiency virus
Hodgkinson, Neville, 119
Hollywood: AIDS and, 113; nuclear power and, 20, 21
homeopathy, 62
homology, evolutionary theory and, 220–26
Hooker Chemical, 64
hormesis: chemical, 57–58, 59; dioxin and, 65; environmental toxicology and, 57, 59; EPA and, 57; radiation

and, 39–46, 51, 53, 57–63. *See also* nuclear power; radiation
Houk, Vernon, 67
Howard Hughes Medical Institute, 141
Howe, Kenneth, 58–59
How to Win a Nobel Prize (Bishop), 170, 171
Huber, Peter, 25
Hudson River, 57
Hueppe, Ferdinand, 61
Human Genome Project, x, 147–48, 157–59. *See also* genetic engineering
Human Genome Sciences, 155
human immunodeficiency virus (HIV), ix, 105–7, 113–14. *See also* AIDS
Hume, David, 183
Huxley, Julian, 199, 206
Huxley, Thomas Henry, 187–88, 206
hydroelectricity, 34

Icons of Evolution (Wells), 225–26
Illmensee, Karl, 124–25, 128
India: DDT and, 82; Kyoto Protocol and, 4; malaria in, 74
Indian Point nuclear power plant, 28
Infectious AIDS: Have We Been Misled? (Duesberg), 179
In His Image: The Cloning of Man (Rorvik), 124
Innis, Roy, 80
In Search of Deep Time (Gee), 223
intelligent design: Catholic Church and, 199; evolutionary theory vs., 199–205, 220
Intergovernmental Panel on Climate Change (IPCC), 6
International Panel on Climate Change, 12
Inventing the AIDS Virus (Duesberg), 119
Inventing the Flat Earth (Russell), 182, 183
Irving, Washington, 183–84

Isacson, Ole, 131–32, 134
It Ain't Necessarily So: The Dream of the Human Genome and Other Illusions (Lewontin), 162

Japan, 55, 143
Jaworowski, Zbigniew, 47
J. Craig Venter Science Foundation, 162
Johns Hopkins, 47, 136, 179
Johnson, Phillip E., 200–201, 225, 230–31, 232
Journal of American Medical Association, 66
Journal of Environmental Monitoring, 61
Journal of Molecular Evolution, 214
Jukes, Thomas, 77
Juvenile Diabetes Research Foundation, 140, 141

Kaczynski, Theodore, 16
Kennedy, Donald, 204
Kennedy, John F., 44
Kennedy, Robert, Jr., 31
Kepler, Johannes, 195, 197
Kettlewell, Bernard, 227
King, Alexander, 76
Klinghoffer, David, 221
Koch, Robert, 61
Koenig, Harold M., 83, 84–85
Kolata, Gina, 42, 124, 125–27, 128, 134, 171
Koop, C. Everett, 113
Kraemer, Duane, 127
Kristof, Nicholas D., 17, 82
Kyoto Protocol: economic impact of, 1; enforcement of, 5; global warming and, 1, 4–5, 10, 26; greenhouse emissions and, 4–5; United States and, 4

Lamarck, Jean-Baptiste, 205
Lammerding, Kurt, 118
Lander, Eric, 149, 156

Lanza, Robert, 141
Lavelle, Rita, 65
Lawrence Livermore National Laboratory, 41
Leaf, Clifton, 166
Lengauer, Christoph, 179
Level 4: Virus Hunters of the CDC (McCormick), 112, 120
Levin, Ira, 123
Lewis, Edward, 43, 44
Lewontin, Richard, 160, 162, 209
"The Life Sciences" (National Academy of Sciences), 75
Lime, Harry, 61
Lindberg, David, 189, 190
Linnean Society, 231
Liroff, Richard, 82
Little Commentary (Copernicus), 192
Little Ice Age, 7, 8, 13–14
Lochbaum, David, 29
Lomborg, Bjørn, 93–94, 100
Loux, Bob, 32
Lovelock, James, 36
Luan, Y. C., 54

MacArthur, Robert, 95, 96
Malan, Rian, 105, 106, 108, 114, 116, 117, 118
malaria: in Africa, 79–81; DDT and, 73–74, 79–81, 83; in United States, 79, 80, 85
Malthus, Thomas Robert, ix, 209
Mancuso, Charlotte, 132
Manhattan Project, 41, 46
Mann, Charles C., 97
Mann, Michael, 7, 9, 10–11
Martin, Gail, 135
Martin, Rowan, 96–97
Marx, Jean, 171
Marx, Karl, vii
Massachusetts–Amherst, University of, 58, 59
Massachusetts General Hospital, 41

Mayr, Ernst, 220, 222
McCain, John, 36
McCloskey, Michael, 83–84
McCormick, Joseph B., 111–12, 120
McIntyre, Stephen, 9–10
McKay, Ronald, 130, 131–32, 139
McKitrick, Ross, 10
media: AIDS and, 105, 110; cancer and, 166–67; DDT and, 73; global warming and, 9, 11; government and, vi; nuclear power and, 20–21; radiation and, 50; science and, v–vii; stem cell research and, xi
Medieval Warm Period, 7, 9
Melton, Douglas, 140
Mendel, Gregor, 190
Mengele, Josef, 123, 187
Meyer, Steven, 221
military: DDT and, 73; nuclear power and, 20, 22, 46
Miller, Kenneth R., 201–2
Miller, Stanley, 228–29
Mills, Mark, 25
Moeller, Dade, 42
The Molecular Biology of the Cell (Alberts), 229
Moore, James R., 190
Moore, Patrick, 36, 90, 91–92, 95
Moore, Ruth, 188
Morbidity and Mortality Weekly Report, 110, 111
Morgan, Thomas Hunt, 207, 212, 215
Morison, Samuel Eliot, 184
"Mother of All Cells" (Cookson), 135
Mueller, Paul, 73
Muller, Hermann J., 232–33
Myers, Norman, 93
"The Myth of Plutonium Toxicity" (Cohen), 50, 51

Nader, Ralph, 25, 50–51, 64, 73
Nagasaki, Japan, 28, 39, 41, 43
National Academy of Sciences, 75, 230

National Cancer Act of 1971, 166
National Cancer Institute, xii, 165, 166, 179, 238
National Council on Radiation Protection and Measurement, 53
National Geographic, 91, 92, 133
National Human Genome Research Institute, 148
National Institutes of Health (NIH), vii, xi, xiii, 130, 237; budget of, vi; cancer and, 163; dioxin and, 69; genetic engineering and, 152; science and media and, v; stem cell research and, 139
National Public Radio (NPR), 9, 150
National Science Foundation, 239
National Toxicology Program, 69
natural selection, evolutionary theory and, 199, 205–6, 207–9
Nature, 10, 12, 15, 59, 96, 125, 149, 152
Nature Conservancy, 96
NatureServe, 96
Naval Nuclear Propulsion Program, 46
Nelson, Gary, 216
Nevada Review Journal, 32
Newman, Edwin, 20–21
New Republic, xiii, 203
New Scientist, 90
Newsweek, 1, 66, 124
Newton, Isaac, 22–23, 149, 191, 197
New Yorker, 74–75, 136
New York Times, vi, 17, 40, 41, 43, 48, 63, 66, 67, 73, 82, 83, 103, 110, 115, 118, 119, 120, 124, 126, 134, 137, 139, 145, 149, 152, 160, 165, 170, 199, 201, 216, 237
Nilsson, Dan E., 211
Nixon, Richard, 165, 166
Nordhaus, Ted, 16–17
North American Free Trade Agreement (NAFTA), 82
Nuclear Energy (Ott and Spinard), 51
nuclear power: alchemy of, 22–25; cancer and, 32; Chernobyl accident and,

28, 47; development of, 19–20; EPA and, 32; fear of, 20–21, 25, 26–29; global warming and, 19, 36–37; "green" civil war and, 29–33; Hollywood and, 20, 21; media and, 20–21; military and, 20, 22, 46; nuclear weapons vs., 22; opposition to, 21–22; terrorism and, 29; United States and, 20; wind turbines and, 27, 30–33. *See also* radiation
Nuclear Regulatory Commission, 50
nuclear weapons, nuclear power vs., 22
Numbers, Ronald, 189, 190

Oak Ridge National Laboratory, 55
Observer (London), 156
Olivier, Lawrence, 123
oncogenes. *See* cancer
Oncogenes, Aneuploidy and AIDS (Duesberg), 119
O'Neill, Paul, 84
One Renegade Cell (Weinberg), 174
On the Origin of Species (Darwin), 89, 187, 205, 213, 215, 220, 227, 229–31, 232, 235
On the Revolutions of the Celestial Orbs (Copernicus), 192
Oreskes, Naomi, 14
Oresme, Nicholas, 183
Owen, Richard, 220

Parkinson's disease, 131; stem cell research and, 131–34, 142
Parkinson's Information Exchange Network, 132
Patterson, Colin, 216–18, 222, 223
Patterson, H. Wade, 45
Pauling, Linus, 43, 44, 45
Peiser, Benny, 14–16
Peterson, Anne, 81
photovoltaic cells, 34
Physics Today, 47
Piedrahita, Jorge A., 127

Pimm, Stuart, 91
Pinker, Steven, 203
Plummer, Mark L., 97
plutonium, 25; toxicity of, 50–54
Politicizing Science: The Alchemy of Policymaking (Gough), xiii
politics: environmentalism and, 89; science and, v–xiii, 14, 28, 75–77, 82–83, 92–97, 105, 180, 237–43; stem cell research and, 133–34
Pollack, Andrew, 155–56
Pombo, Richard, 103
Popper, Sir Karl, 208
population: of Africa, ix, 100, 105; AIDS and, 118–19; biodiversity and, 88; over-, ix
Porter, John Edward, 239–40
Poultry Science, 78
"The Power to Divide" (Weiss), 133
Principles of Political Economy (Malthus), 209
Provine, William, 200, 201
psychology, evolutionary, 88
public policy: DDT and, 83; radiation and, 39, 43

Quammen, David, 94–95

Racing to the Beginning of the Road: The Search for the Origin of Cancer (Weinberg), 168
radiation: background, 19–21, 24, 46–48, 51, 52; beneficial effects of, 61–63; birth defects and, 42; cancer and, 39, 42, 43–44, 48, 51, 52, 60, 63, 168; Chernobyl accident and, 48–50; coal vs., 46; dioxin and, 63–65; environmentalists and, 39; EPA and, 39, 46; fallout and, 43–44; fear of, 45; gamma rays and, 54–55; half-life and, 24, 46; hormesis and, 39–46, 51, 53, 57–63; linear no-threshold theory of, 40, 43–45; media and, 50;

radiation (continued):
   natural, 25, 47, 51, 52; nuclear power
   and, 22–25; plutonium toxicity and,
   50–54; public policy and, 39, 43;
   radon and, 39, 51–53, 55; therapeutic, 55; UN and, 49. *See also* nuclear
   power
"Radiation Risk and Ethics" (Jaworowski), 47
radon, 39, 51–53, 55. *See also* radiation
The Reactor Shielding Design Manual
   (Rockwell), 46
religion: evolutionary theory and, 216;
   faith and, xiii; science and,181–97,
   204
Religion and Science (Russell), 189–91
renewable energy: environmentalists
   and, 27, 29–30, 34–37; problems
   with, 35; tax credit for, 35
Rennie, John, 135
Richardson, Michael, 226
Rickover, Hyman, 46
Ridley, Matt, 94
Ried, Thomas, 179
Ripley, S. Dillon, 92
Roberts, Donald, 83
Rockwell, Theodore, 46, 49–50
Roll Back Malaria, 73
Rolling Stone, 105, 106
Romanes, G. J., 190
Rooney, Andy, 119
Rorvik, David, 124
Rosenberg, Tina, 79
Rosenthal, Elisabeth, 49
Rosenthal, Nadia, 141
Roslin Institute, 125
Rous, Peyton, 169, 170, 175
Ruckelshaus, William, 75, 78–79, 81
Russell, Bertrand, 189–91, 209
Russell, Jeffrey Burton, 182, 183, 184, 188

Sagan, Carl, viii, 229
Sagan, Leonard, 59

Salk Institute, 14
Sanera, Michael, 97
Sanger Institute (Britain), 171
Sarton, George, 189
Schaefer, Mark, 96
Scheff, Liam, 105
Schlesinger, James, 10
Schneider, Stephen, 2
Schoofs, Mark, 110
Schulz, Hugo, 61, 62
Schwann, Theodor, 143
Schwegman, A. B., 118
Schweitzer, Albert, 44
SCID: severe combined immunodeficiency, 159–61, 162
science: challenging, v–vii; climate,
   viii, 1–17; competition of theories
   and, xii; consensus and, xiii, 13–14;
   crisis and, ix; economics and, viii;
   ethics and, xi; fashion and, 39; flat
   earth myth and, 182–86; funding
   and, xii–xiii, 16; media and, v–vii;
   medical, vi, vii; national security
   and, vi; politicization of, v–xiii, 14,
   28, 75–77, 82–83, 92–97, 105, 180,
   237–43; private-sector research and,
   xii; religion and, 181–97, 204; uncertainty in, xiii
Science magazine, xii–xiii, 8, 9, 14, 15,
   44, 60, 76, 77–78, 101, 110, 125, 132,
   134, 138, 141, 142, 149, 151, 171,
   204, 222, 226, 239
Science, Money, and Politics (Greenberg), 237
Science Under Siege (Fumento), 64
Scientific American, 14, 63, 94, 96,
   131, 135, 137, 141, 148, 149, 178,
   227, 233
Scientific Committee on the Effects of
   Atomic Radiation, 49
Seveso, Italy, 64, 65, 66, 70–71
Shapin, Steven, 190
Shaw, Jane S., 97

Shaywitz, David A., 143
Shellenberger, Michael, 16–17
"Show Me the Science" (Dennett), 211
Shreeve, James, 162
Shute, Neville, 20
sickle-cell anemia, 157, 158
Sierra Club, 16, 83–84
Silent Spring (Carson), 74–75, 76, 79–80
Simon, Julian, 93
Singer, Fred, 5, 12–13, 15
The Sinking Ark (Myers), 93
The Skeptical Environmentalist (Lomborg), 93, 100
Slobodkin, Lawrence, 96
Smith, Fred, 100, 102
Smith, J. D., 223
Smith, John, 50
Smithsonian Institution, 88, 92
Sociobiology: The New Synthesis (Wilson), 88
The Solar Fraud (Hayden), 25, 26
Solter, Davor, 125
Soon, Willie, 8
Soviet Union, 145; Kyoto Protocol and, 4; radiation and, 47, 49
Specter, Arlen, 241
Specter, Michael, 121
Spencer, Herbert, 206
Spinard, Bernard I., 51
Stanford University, 83, 136, 137
Stanley, Wendell, 168, 175
State Department, vi, ix, 73, 121
State Nuclear Projects Agency, 32
State of Fear (Crichton), 13
Stein, Rob, 171
stem cell research, 123; Alzheimer's disease and, 134, 136, 140; cancer and, 131, 135; cloning and, 129–30; diabetes and, 136, 140–42; government funding for, x, 131, 142–43; media and, xi; Parkinson's disease and, 142; politics and, 133–34; problems with, 137–45; science and religion and, 204; success of, 131
Sternberg, Richard, 221
Stewart, Christine, 5
Stockholm University, 11–12
Stokes, Louis, 239–40
Stokes, Sir George, 190
Strauss, Lewis L., 20, 43
Stroup, Richard L., 93
Sunday Times (London), 50
Superfund legislation, 63
Sweeney, Edmund, 78

Taiwan, 53–54
Tatum, Edward, 151
Teaching About Evolution and the Nature of Science, 230
Teller, Edward, 44, 45
Texas Heart Institute, 137
Thomson, James A., 137–38
Three Mile Island, Pa., 21
Time magazine, 28, 29, 124, 201, 203
Times Beach, Mo., 57, 65, 67
Todaro, George, 169
toxicology: animal, 69–70; environmental, 57, 57–71
Toxic Substances and Disease Registry, 69–70
"The Tragedy of the Commons" (Hardin), 101

Ukraine, 47, 49
UNAIDS, 105
Uncommon Dissent: Intellectuals Who Find Darwinism Unconvincing (Dembski), 214
Uniformed Services University of the Health Sciences, 83
Union of Concerned Scientists, 28
United Nations (UN): African AIDS and, 105; Chernobyl accident and, 48; global warming and, 6; nuclear power and, 19; radiation and, 49

UN Security Council, ix
United States: AIDS in, 106, 108; Kyoto
    Protocol and, 4; malaria in, 79, 80,
    85; nuclear power and, 20
U.S. Agency for International Develop-
    ment (USAID), 81, 118, 119
U.S. Centers for Disease Control (CDC),
    65, 67, 106, 107, 110, 111, 113
U.S. Fish and Wildlife Service, 33, 74,
    77, 102
U.S. News & World Report, 60
U.S. Public Health Service, vii, 73, 111
uranium. See nuclear power; radiation
Urey, Harold, 228–29

Varmus, Harold, 169–71, 173, 178, 239,
    240
Venter, J. Craig, genetic engineering
    and, 148–52, 155, 156, 161–63
Vietnam War, vi, 64
Village Voice, 110
Vogelstein, Bert, 179
von Storch, Hans, 11–12
Vorilhon, Claude, 124

Waddington, C. H., 206
Wade, Nicholas, 139–40
Wald, George, 45
Wald, Matthew, 40
Wallace, Alfred Russel, 208, 231
Wall Street Journal, 10, 11, 73, 82, 110,
    140, 141, 158, 161, 221, 230–31
The Warfare of the Sciences (White),
    184
Warrick, Joby, 42
Washington Post, vii, xi, 33, 34, 42, 63,
    66, 127, 136, 138, 143, 149, 150, 159,
    161, 171
Watson, James, x, 134, 149, 154, 156,
    161
Watson, Lyall, 223

Waxman, Henry, 238
Webster, Edward, 41
The Wedge of Truth: Splitting the Foun-
    dations of Naturalism (Johnson), 232
Weekly Epidemiological Record, 110,
    113–14
Weikhart, Richard, 214
Weinberg, Robert, 168, 173, 174, 176
Weiss, Rick, 126, 127, 133, 161
Weissman, Irving, 136, 137
Wells, Jonathan, 222, 223, 225–26, 234,
    234–35
West, Michael, 128, 143
White, Andrew Dickson, 184–87, 189
Whitehead Institute, 149, 156
"Why I Am an Infidel" (Burbank),
    233–34
Wilberforce, Bishop, 187–88
Wildavsky, Aaron, 93
Wildlife Conservation Fund, 98
Wilkins, John, 191
Wilmut, Ian, 125, 127
Wilson, Edward O., 87–89, 92, 94–96,
    97, 99, 162
Wolfensohn, James, 118–19
Woodward, Bob, vii, xi, 243
World Bank, 84, 118
World Health Organization (WHO), 74,
    77, 81, 107, 110, 112–14
World Trade Organization (WTO), 88
World Wildlife Fund, 82, 95

X-rays, 62, 167–68, 232

Yes to Human Cloning: Immortality
    Thanks to Science! (Vorilhon),
    123–24
Yucca Mountain, Nev., 32
Yushchenko, Viktor, 69

Zwiers, Francis, 11